MECHANICAL FACE SEAL HANDBOOK

MECHANICAL FACE SEAL HANDBOOK

JOHN C. DAHLHEIMER

CHILTON BOOK COMPANY
PHILADELPHIA NEW YORK LONDON

Published in Philadelphia by Chilton Book Company
and simultaneously in Ontario, Canada,
by Thomas Nelson & Sons Ltd.
Designed by Rosanna V. de Plata
Manufactured in the United States of America

Library of Congress Cataloging in Publication Data

Dahlheimer, John C. 1940–
Mechanical face seal handbook.

1. Sealing (Technology) I. Title. II. Title:
Face seal handbook.
TJ246.D34 621.8′85 72-6443
ISBN 0-8019-5747-8

PREFACE

In spite of the large and ever growing usage of mechanical face seals, standard engineering texts and handbooks contain little or no information on this most versatile type of sealing device. The purpose of this handbook is to provide a comprehensive basic reference work written expressly for users of mechanical face seals . . . present and potential.

Regarding terminology, the name of the basic type sealing device with which this handbook deals has, in the past, defied standardization. The following terms have all been used to describe this one specific type of seal: face seal, axial mechanical seal, end-face seal, mechanical shaft seal, mechanical seal, rotary mechanical seal, rotary face seal and mechanical face seal. In recent years, however, "mechanical face seal" has become by far the most widely and commonly used term in technical literature dealing with this type of seal, undoubtedly because it is the most historically comprehensive, least ambiguous and, therefore, most useful term of all. Since common usage is the key to standardization, it would seem at long last the industrial and scientific communities at large have given de facto acceptance to "mechanical face seal" as the standard term to use when talking or writing about this type of seal.

Many of the devices, materials, and methods of manufacture mentioned or illustrated in this handbook are currently protected by patents in the United States and other nations, or have patents pending and their inclusion does not constitute a recommendation for their unauthorized use.

John C. Dahlheimer

ACKNOWLEDGMENTS

We acknowledge proprietary rights, as indicated, of the following names and designations used in this text:

Surfindicator—Brush Instrument Div.
Polaroid—Polaroid Corp.
Instron—Instron Corp.
Teflon—E. I. du Pont de Nemours and Co., Inc.
Ni-Resist—International Nickel Co.
Monel—International Nickel Co.
Stellite—Stellite Div., Cabot Corp.
Ni-Span-C—International Nickel Co.
Rene 41—Allvac Co.
Hastelloy—Stellite Div., Cabot Corp.
Wallace—Testing Machines, Inc.
Durometer—The Shore Instrument and Mfg. Co., Inc.
Freon—E. I. du Pont de Nemours and Co., Inc.
Elgiloy—Elgin National Watch Co.
Havar—Hamilton Watch Co.
Sandvik—Sandvik Steel Co.
Berylco—Beryllium Corp.
Duranickel—International Nickel Co.
Permanickel—International Nickel Co.
LW1—Linde Co., Div. of Union Carbide Corp.
LC-1A—Linde Co., Div. of Union Carbide Corp.
LA-2—Linde Co., Div. of Union Carbide Corp.
S-816—Allegheny Ludlum Steel Co.
L-605—Allvac Co.

(Any omissions from this list are unintentional.)

CONTENTS

ILLUSTRATIONS AND TABLES

MECHANICAL FACE SEAL HANDBOOK

Mechanical Face Seals—The most versatile of all seals for rotating shafts.

Exploded View of a Typical Mechanical Face Seal.

1
INTRODUCTION

A. DESCRIPTION

A mechanical face seal is a device used to effect a pressure-tight seal between a rotating shaft and a member through which the shaft passes. Sealing is performed by continuous relative contact between two ultra-flat radial sealing faces located in a plane perpendicular to the shaft centerline. One face is attached to the shaft and rotates with it; the other to the housing, which is stationary. One face is designed to move axially to allow contact between the two at all times, with a preload force normally used to insure this contact.

In addition, two secondary and relatively static seals are used to keep the medium being sealed from escaping around the two sealing faces and the respective members to which they are attached. The secondary seals are usually made of an elastomeric material and take the form of O-rings, square section rings, bellows, gaskets and so forth.

In some instances, one of the sealing face members is mounted onto the shaft, or into a bore of the housing surrounding the shaft, by means of a press-fit, thus eliminating the need for an elastomeric static seal for that member. Or one of the sealing faces may be provided on an integral part of the mechanism in which the seal is installed. For example, lapping the end of a bearing race, stuffing box flange, or an impellor hub surface to a precise flatness and finish provides a surface against which the seal may run.

B. APPLICATIONS

Mechanical face seals are usually specified for sealing applications which cannot be handled adequately by radial oil seals and are used in most applications formerly sealed by jam-type packings in stuffing boxes. Mechanical face seals are used in devices such as pumps, motors, appliances—dishwashers, clothes washers, disposals, refrigerators, freezers and air conditioners; engines and their accessories, power transmission equipment, machine tools, process equipment, control valves, rotary unions, even atomic and nuclear reactors.

C. OPERATING CAPABILITIES (design and material dependent)

It is impractical to define the limits of operation of mechanical face seals because conditions must be considered in many combinations. Although special seals can be and have been designed to go beyond the following limitations, in general they are not covered in this handbook.

1. Fluid pressures to 3,000 psi.
2. Shaft speeds to 50,000 fpm.
3. Ambient temperatures from −425°F to +1200°F.

Cut-Away View of Automotive Water Pump—Showing the mechanical face seal assembly between the pump impellor and bearing.

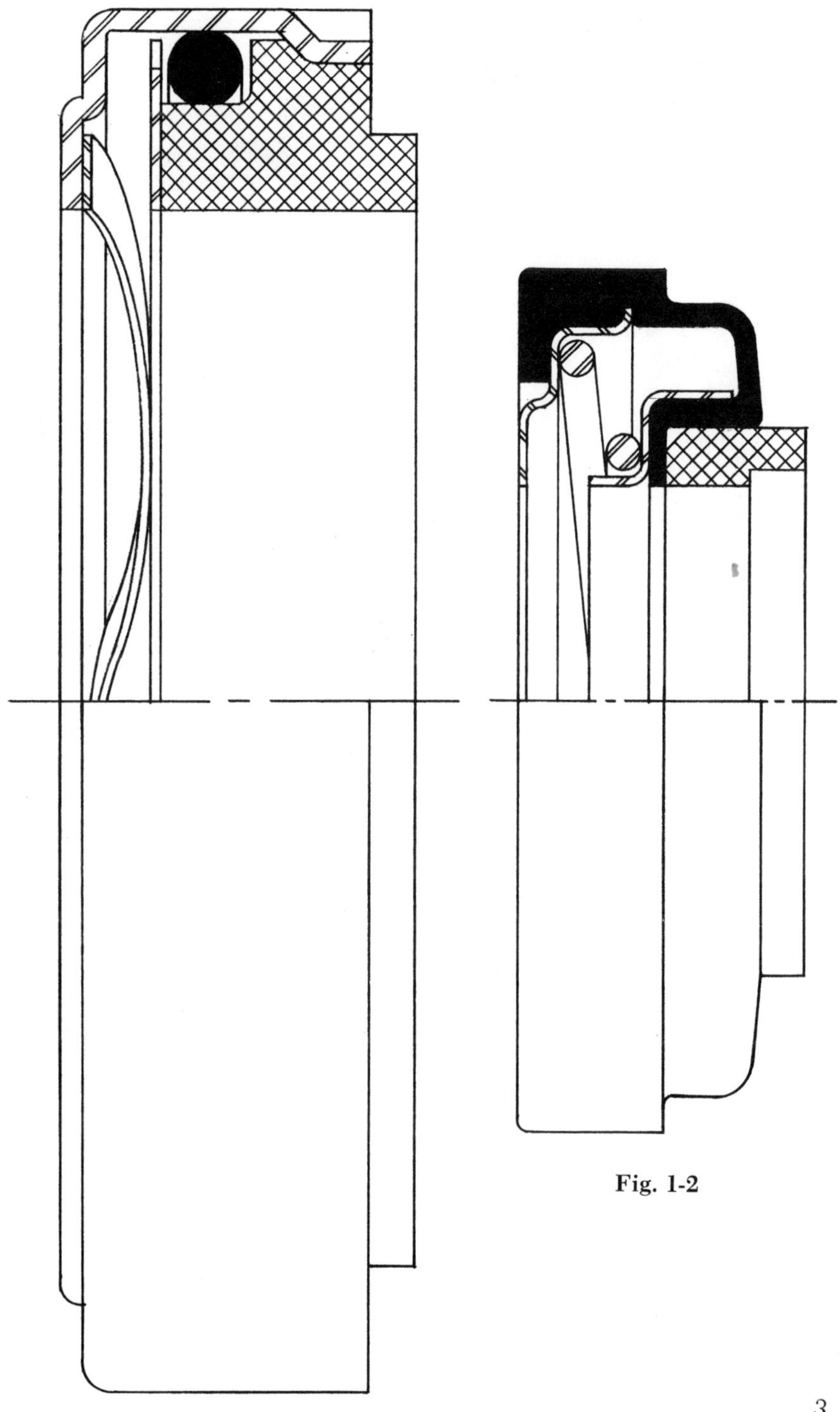

Fig. 1-1

Fig. 1-2

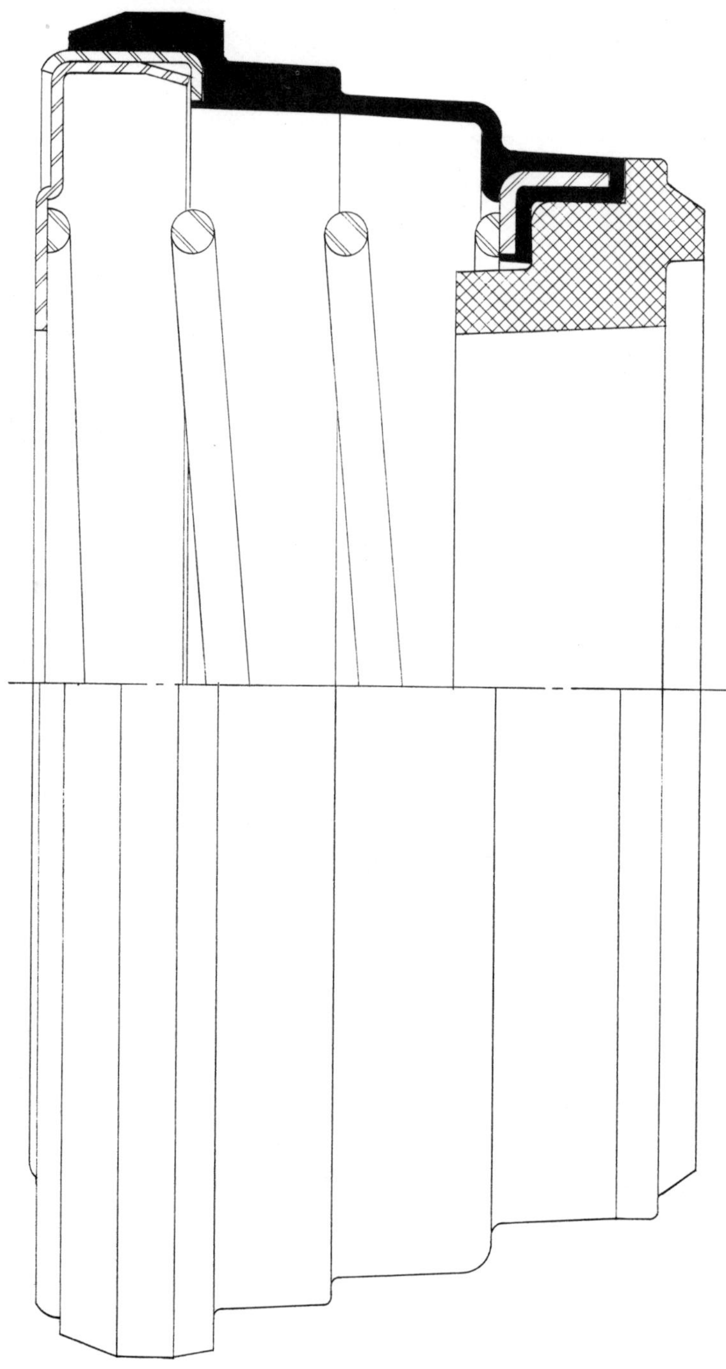

Fig. 1-3

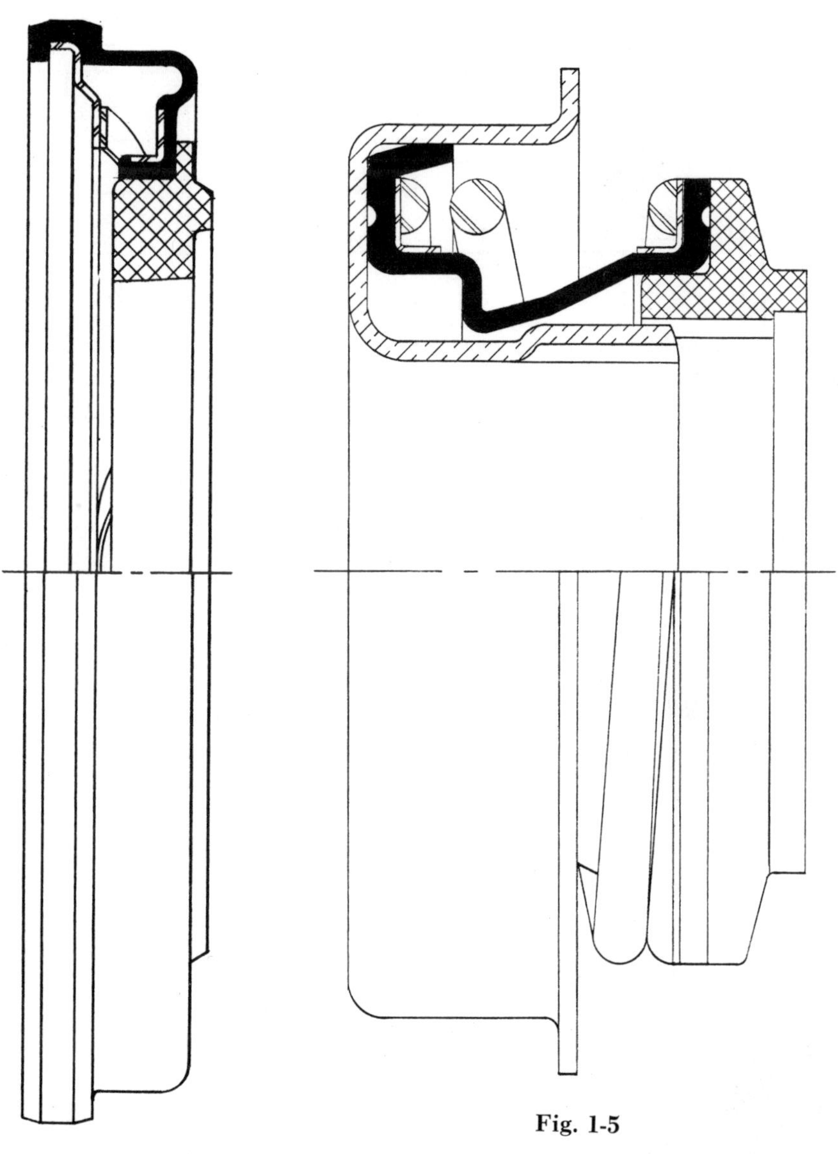

Fig. 1-4

Fig. 1-5

Assorted Seal Head Designs (Cont'd)

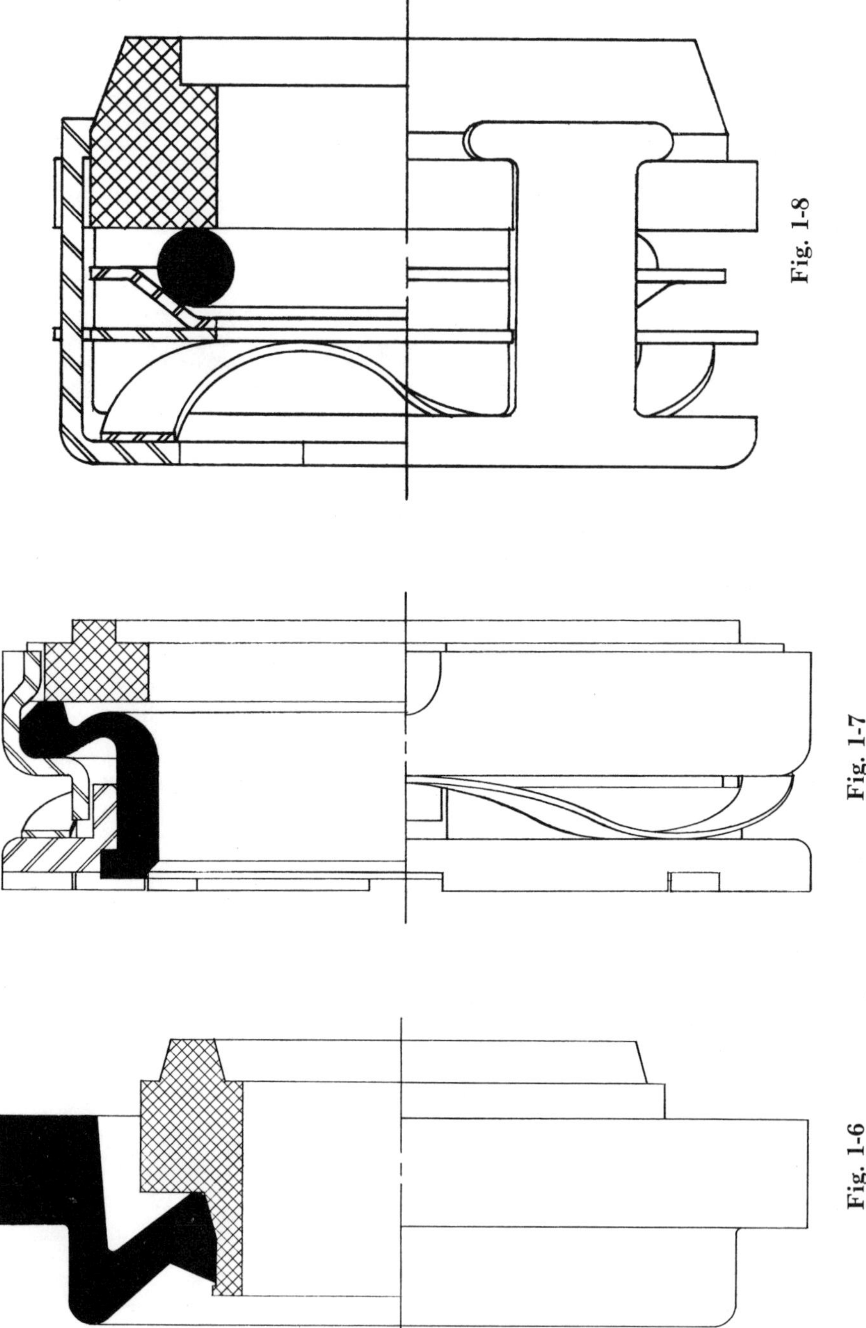

Fig. 1-8

Fig. 1-7

Fig. 1-6

Assorted Seal Head Designs (Cont'd)

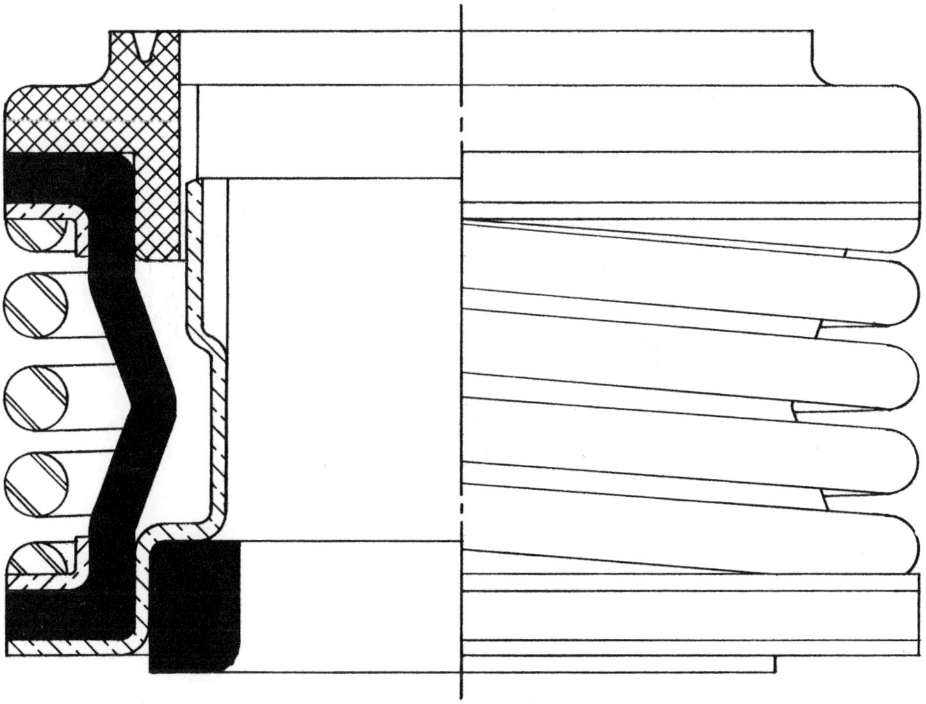

Fig. 1-10

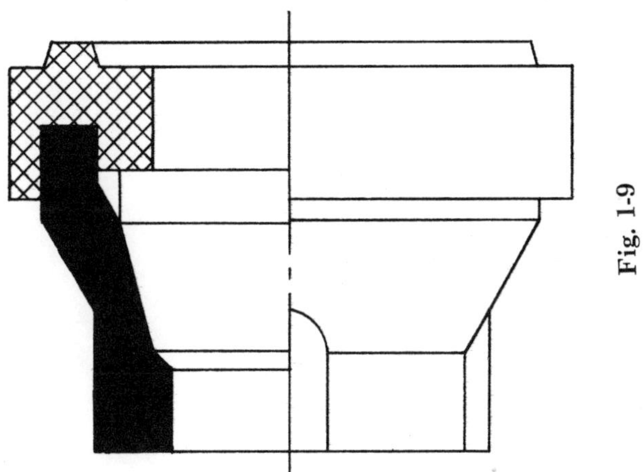

Fig. 1-9

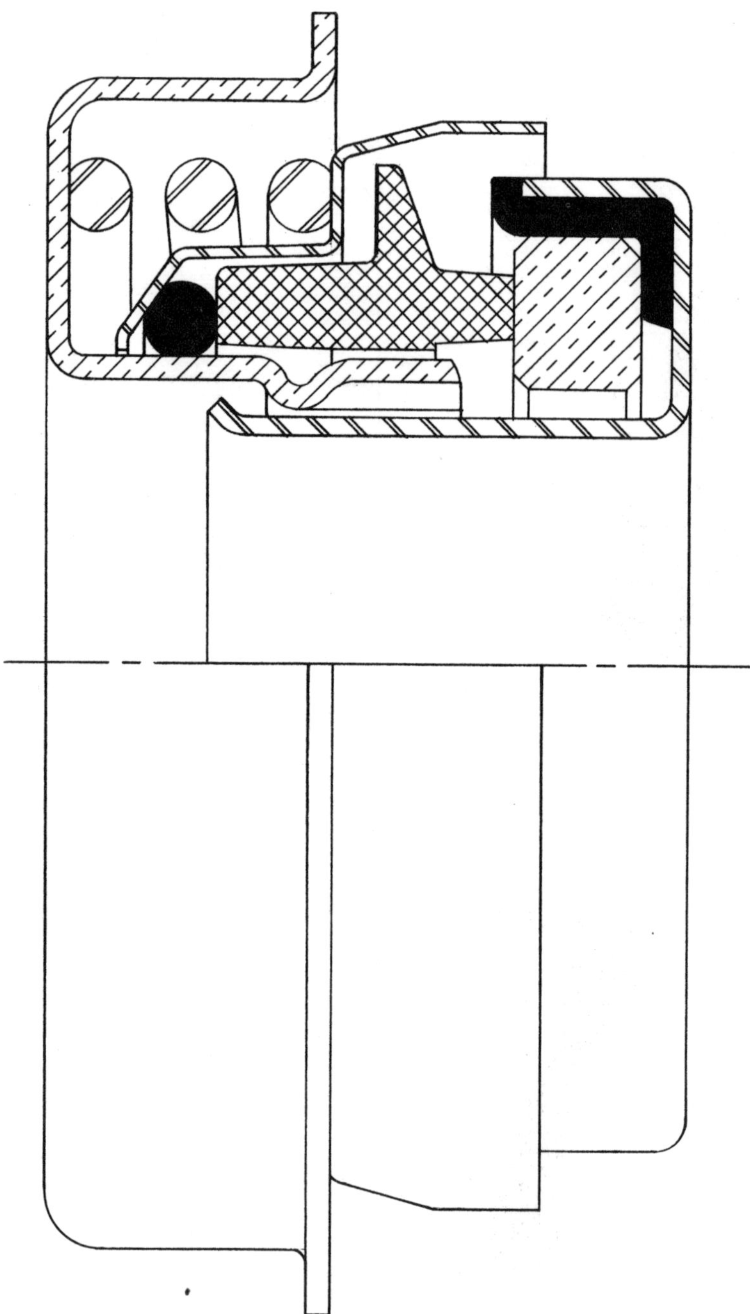

Fig. 1-11

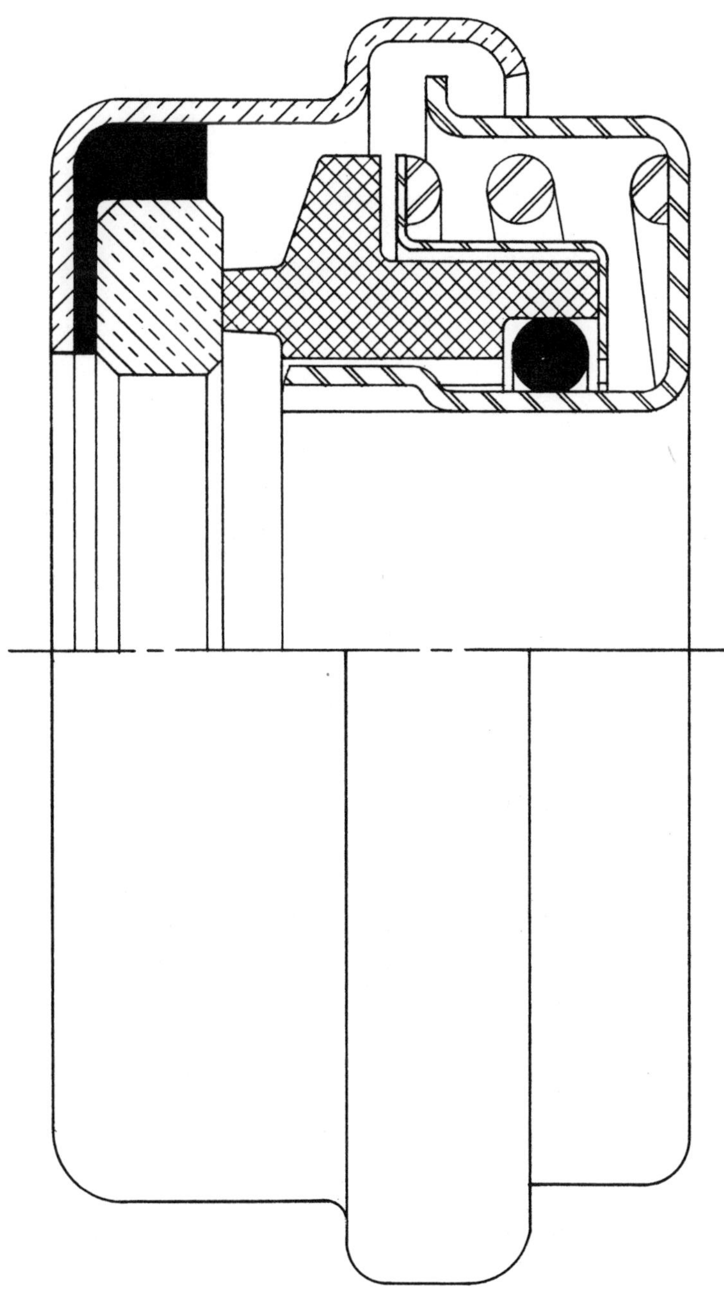

Fig. 1-12

D. ADVANTAGES OF USING

1. Can handle all types of fluids including those containing acids, salts, strong oxidizing and reducing agents, and abrasive particles.
2. Will operate satisfactorily even when shaft and bore are slightly misaligned or non-concentric.
3. Mechanical face seals are unique among all seals in their ability to function positively, both statically and dynamically, with bidirectional shaft rotation, large pressure reversals, and a wide diversity of mediums, temperatures, pressures, and speeds.
4. Shaft finish, roundness, hardness, and material are not highly critical.
5. Operation does not cause shaft wear.
6. Specifying properly designed and manufactured mechanical face seals usually results in reduced warranty and liability cost, reduced downtime, less loss of processing fluids, and longer operating life than experienced with other types of dynamic sealing devices.
7. Additional advantages may be realized in specific applications, such as elimination of system contamination by packing fragments in many "clean" processes, ease of sterilization for food processing installations and reduction of leakage from systems containing hazardous fluids.

E. LIMITATIONS AND ALTERNATIVES

1. *Space.* Mechanical face seals usually require more space than a radial oil seal. Custom-designed mechanical face seals can, however, minimize the space required by as much as 40% as compared to stock or off-the-shelf designs.
2. *Axial shaft movement.* Conventional mechanical face seals are usually not designed to accommodate more than 1/16 inch total axial variation in installed or operating heights. Custom-designed mechanical face seals can be made, however, which accommodate 3/8 inch total axial movement or more. (See Fig. 1-3.)
3. *Handling.* In order to create and maintain a liquid-tight seal, the sealing faces or surfaces of mechanical face seals are sometimes finished to a flatness of 11.6 microinches and as smooth as 2 microinches. In addition, these types of seals are manufactured to cleanliness levels similar to those in the precision bearing industry. As with most precision devices, rough or careless handling must be avoided. However, custom-designed, fully unitized sealing assemblies can be provided in which all components necessary for

Seal Testing Machines—Used by both manufacturers and users of seals for studying leakage rates, wear rates, life expectancy, failure modes, and heat and noise outputs; also for conducting value analysis of seal designs, components, and materials; for quality assurance and qualification testing; and for evaluating manufacturing and installation procedures.

Washing Machine Transmission Test Stand—For performing accelerated life tests on finished mechanical face seals.

Automotive Water Pump Test Stand—For performing accelerated service tests on finished mechanical face seals.

accomplishing the complete sealing function are supplied in a single unit which requires little special attention to handling. (See Figs. 1-11 and 1-12)

4. *Cost.* Mechanical face seals usually cost more initially than radial oil seals and most other common sealing devices; they should therefore be specified only where service conditions, life or leakage requirements, pollution control, safety, or overall engineering and economic considerations make their use appropriate.

F. STOCK *VS* CUSTOM

1. Stock Seals

Stock seals are those offered in the form of a basic design, which can or have been manufactured in many different sizes; they are made available by many mechanical face seal manufacturers as an answer to a fairly wide range of possible operating conditions and customer requirements. Purchases of stock mechanical face seals usually do not require payment of a tooling-up charge since the manufacturer either already has the required tooling or is willing to absorb the cost. When full tooling is already available, delivery times are often shorter than in the case of custom designs for which tooling must be built. The performance of a stock mechanical face seal is usually adequate for most applications even though, by their very nature, they do not always reflect the most current technological advancements, and generally represent engineering compromises in an attempt to accommodate the broadest combinations of operating environments and customer requirements possible. Even though they may be needlessly large, complex, and expensive for a particular application, the use of stock mechanical face seals is still justifiable and recommended when the number of units to be sealed is relatively low and when none of the operating conditions, such as speed, pressure, temperature, leakage allowances, torque, or operational life requirements, are extreme.

2. Custom Seals

Since no two sealing applications are identical, it is often advantageous to use mechanical face seals which are custom designed and produced for a particular application. Instead of designing a mechanism to accommodate the size and operating restrictions of available stock mechanical face seals, for overall engineering and economic reasons it often makes more sense to use a mechanical face seal which has been specifically designed to suit the size, operational, and installation requirements of the particular mechanism. Moreover, when the number of units to be sealed is relatively high one can usually effect a unit cost savings by using a seal that has been value-engineered for one particular use. Additional space and monetary savings can often be realized with custom-designed mechanical face seals which serve dual purposes. Such seals have been used as valves, thrust bearings and bearing housings in many applications. Examples of custom designed and manufactured mechanical face seals are found in most home appliances, auto air conditioning compressors and water pumps, ships, submarines, aircraft, and a host of other applications.

2
DESIGN AND ENGINEERING

A. GENERAL

The apparent simplicity of mechanical face seals belie their inherent sophistication. In their design, the interrelationship of various environmental factors such as speeds, pressures of mediums being sealed, ambient temperatures, vibrations, eccentricities, alignments, and tolerances, must be taken into account, as well as the nature of the medium being sealed and surrounding materials. These factors must be balanced against performance required, space available, materials available, manufacturing technology, production costs, and installation requirements.

B. ROTATING OR STATIONARY SEAL HEAD

In a mechanical face seal, the seal head may normally be defined as that assembly consisting of a housing or case, the seal nose or seal washer, a secondary seal such as an O-ring or bellows, and a preloading device such as a helical-compression spring or wave washer spring.

A mechanical face seal can be designed so that the seal head assembly is mounted on, and rotates with, the shaft (Figs. 2-1, 2-2), or so that the seal head assembly is fixed into a seal bore, with the inside diameter sufficient for clearance from the shaft, which rotates freely through the center of the seal head (Figs. 2-3, 2-4).

The rotating head design best accommodates shaft-center to bore-center offset, or dynamic shaft whip during operation. The sta-

Four Types of Seals With Rotating Heads and Stationary Seats

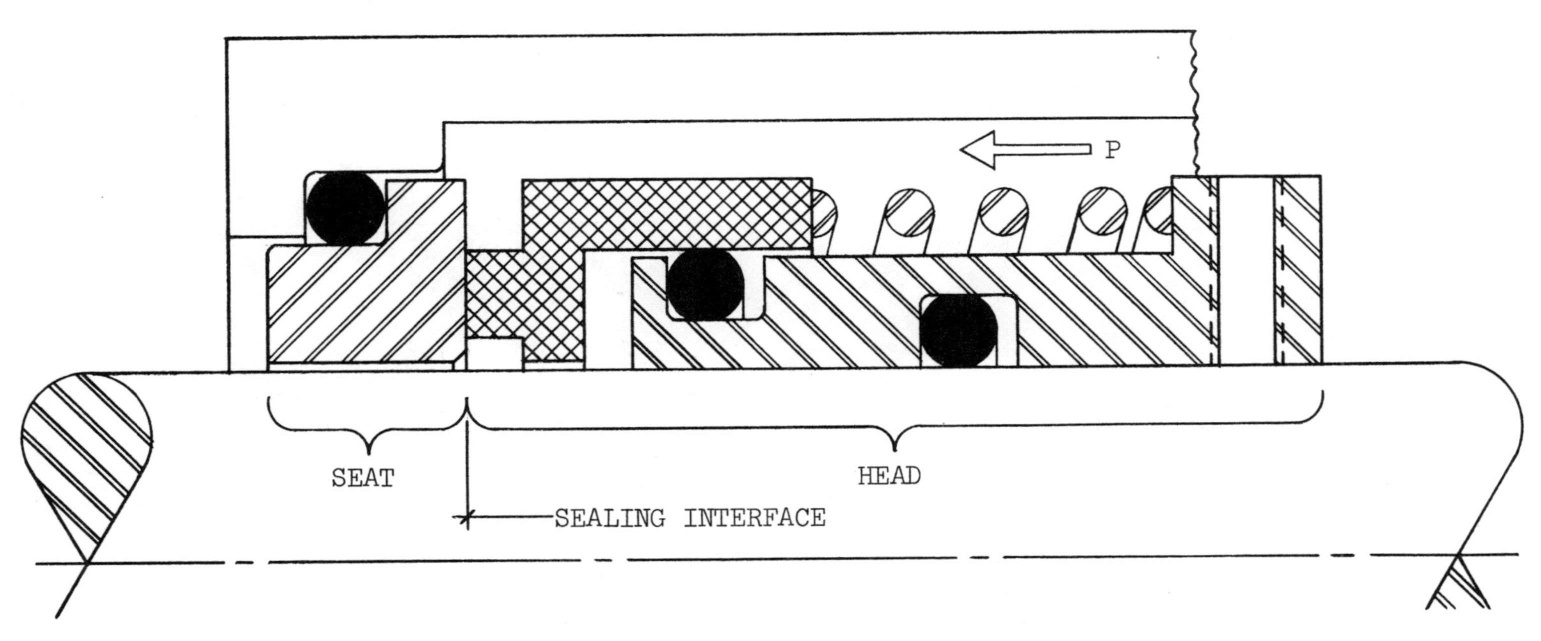

Fig. 2-1a External Pressure, Underbalanced

Four Types of Seals With Rotating Heads and Stationary Seats (Cont'd)

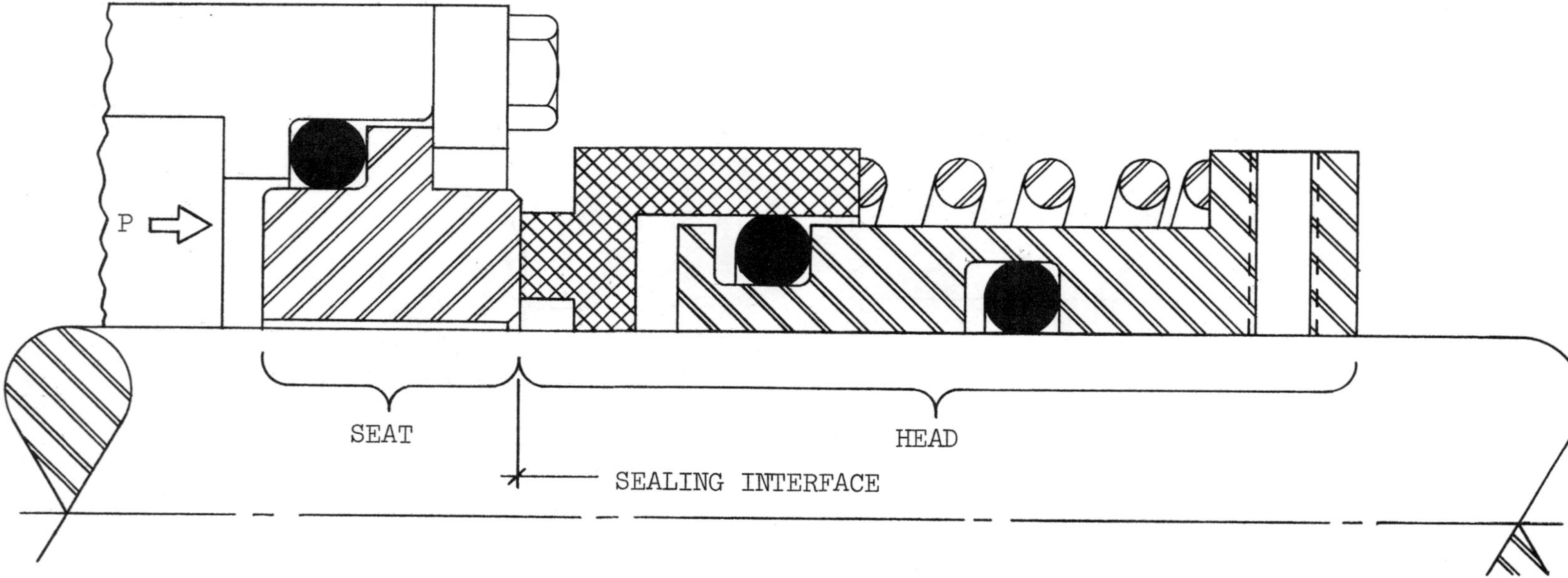

P

SEAT

HEAD

SEALING INTERFACE

Fig. 2-1b Internal Pressure, Overbalanced

tionary head-type seal design, on the other hand, best accommodates shaft-center to bore-center misalignment and high shaft speeds. However, the space available for the entire sealing package, and installation and seal removal considerations, are usually the deciding factors as to which type seal is best suited for a given application.

C. SEAL BALANCE

Mechanical face seals required to operate at high pressure differentials, high relative speeds, elevated temperatures, or in poor lubricants, require that special attention be paid to their hydraulic balance. Hydraulic balance is defined by the mathematical ratio of two areas: usually the ratio of the area of the sealing face which is bounded by the hydraulic balance diameter of the seal assembly and the inside diameter of the seal face, to the area which is bounded by the outer and inner diameters of the sealing face. Proper attention to hydraulic balance allows the seal to run cooler, wear less, and absorb less torque, while keeping the leakage less than the permissible maximum. The type and degree of balancing required varies, depending on the construction of the seal, and the operating conditions under which the seal must function. The viscosity and lubricity of the medium being sealed, the rotational speeds, the location of the sealed medium in respect to the seal, the possibilities of surge pressures or pressure reversals, and other factors, must all be taken into account. A given seal, operating under a given set of conditions may be classified as underbalanced, balanced, or overbalanced.

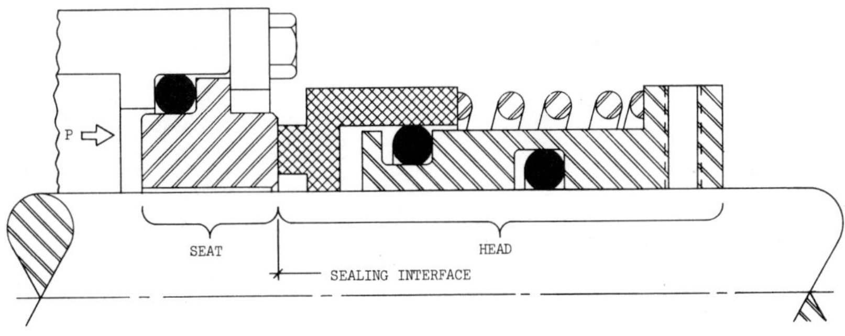

Four Types of Seals With Rotating Heads and Stationary Seats (Cont'd)

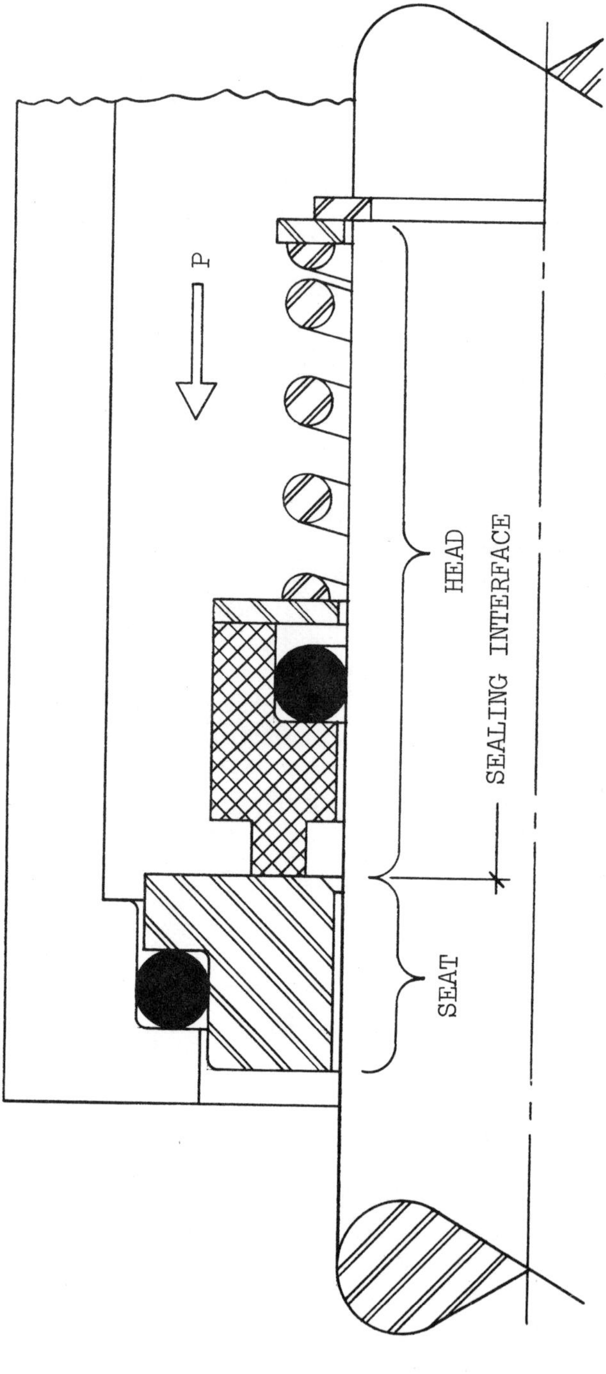

Fig. 2-2a External Pressure, Overbalanced

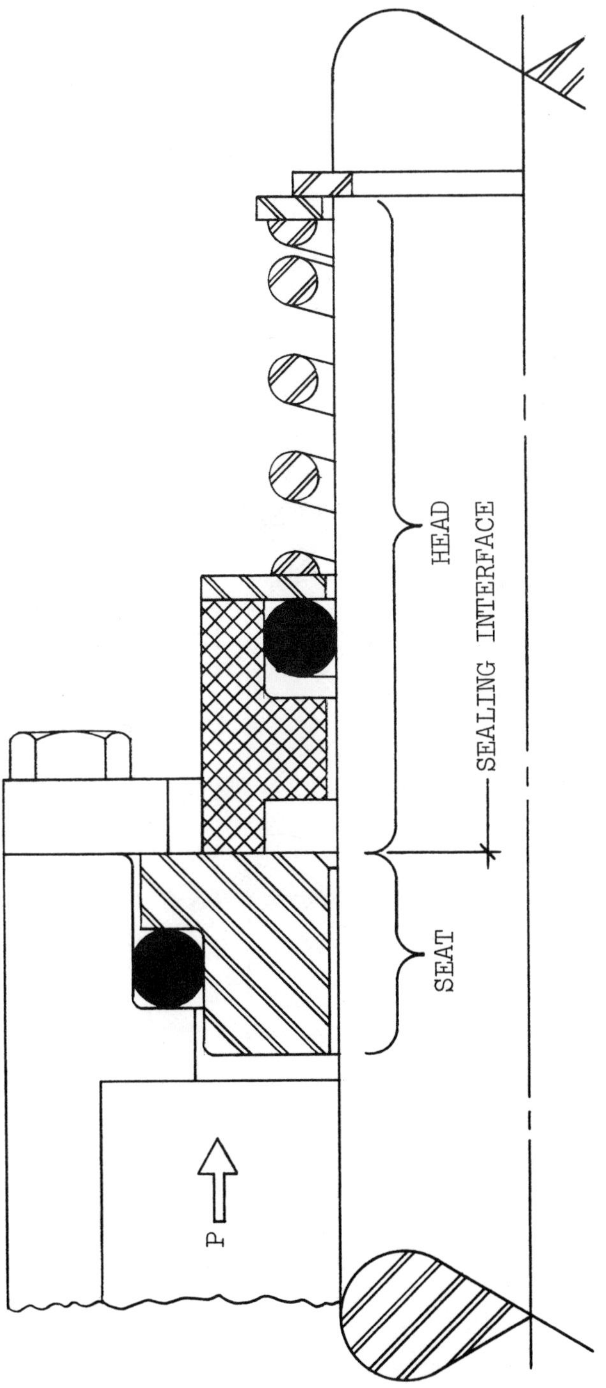

Fig. 2-2b Internal Pressure, Underbalanced

Four Types of Seals With Stationary Heads and Rotating Seats

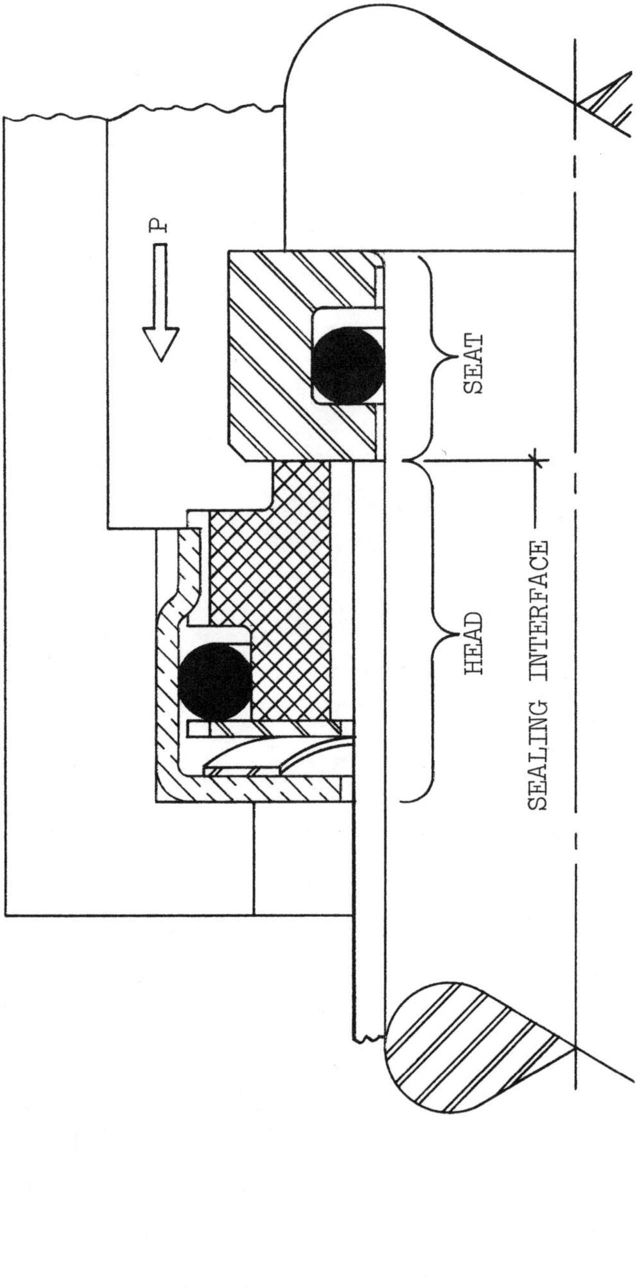

Fig. 2-3a External Pressure, Underbalanced

Four Types of Seals With Stationary Heads and Rotating Seats (Cont'd)

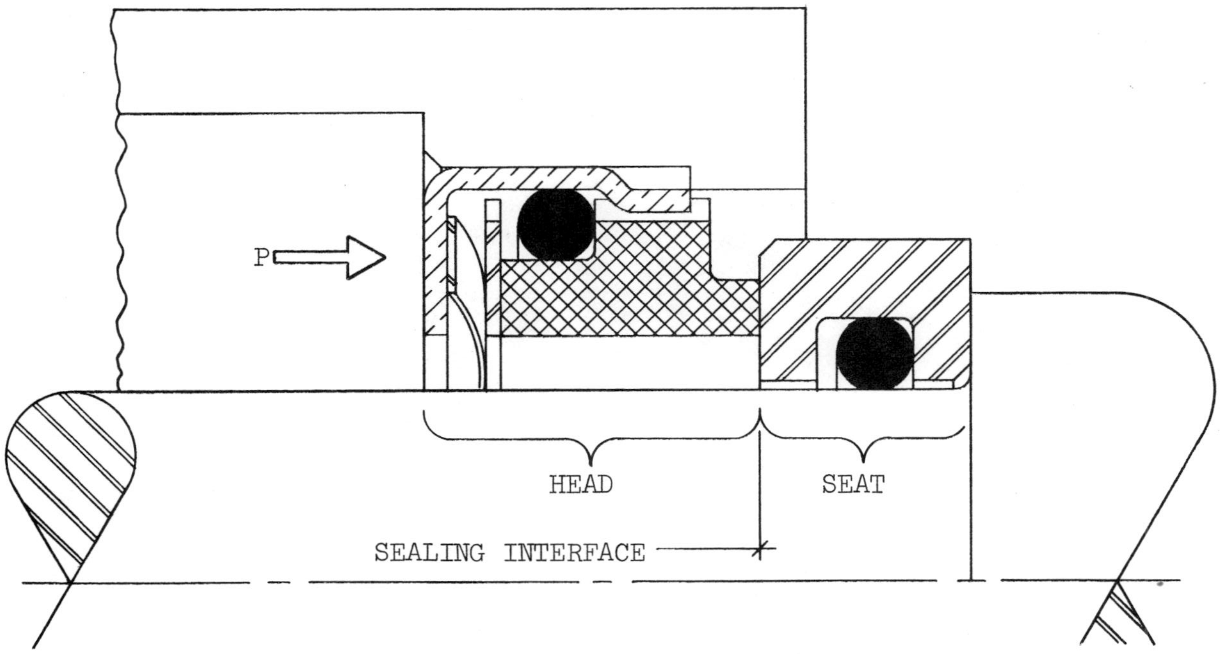

Fig. 2-3b Internal Pressure, Overbalanced

Four Types of Seals With Stationary Heads and Rotating Seats (Cont'd)

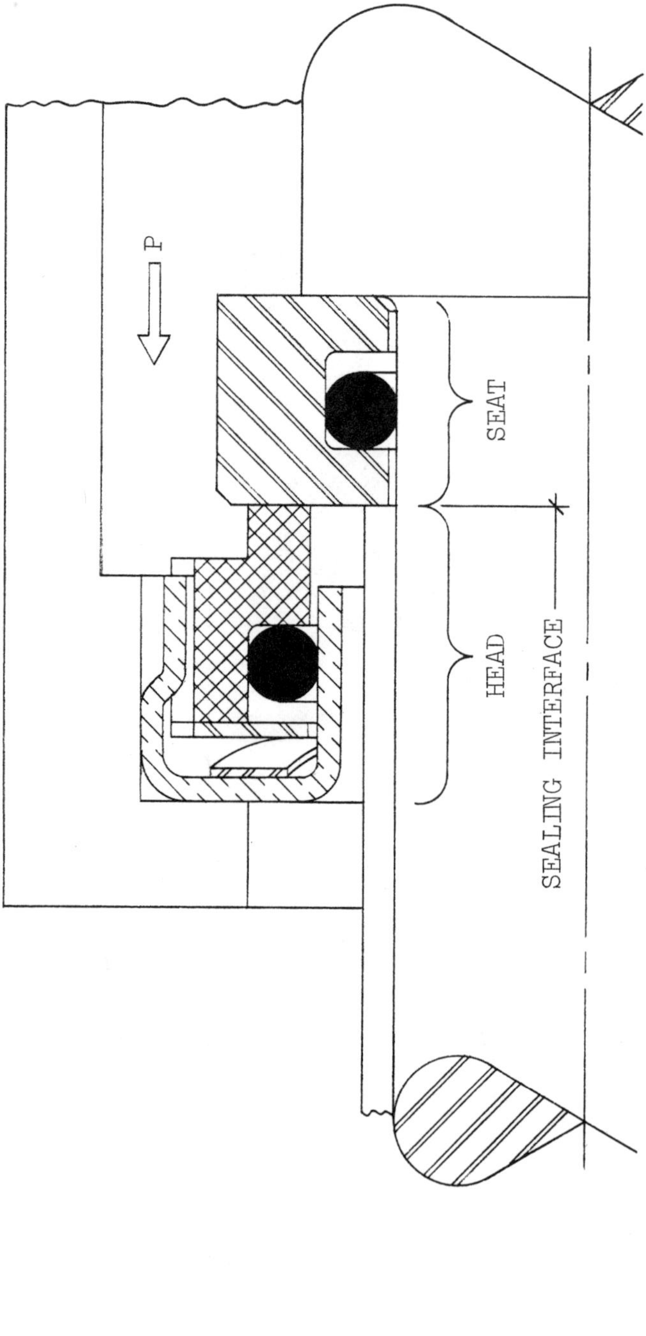

Fig. 2-4a External Pressure, Overbalanced

Four Types of Seals With Stationary Heads and Rotating Seats (Cont'd)

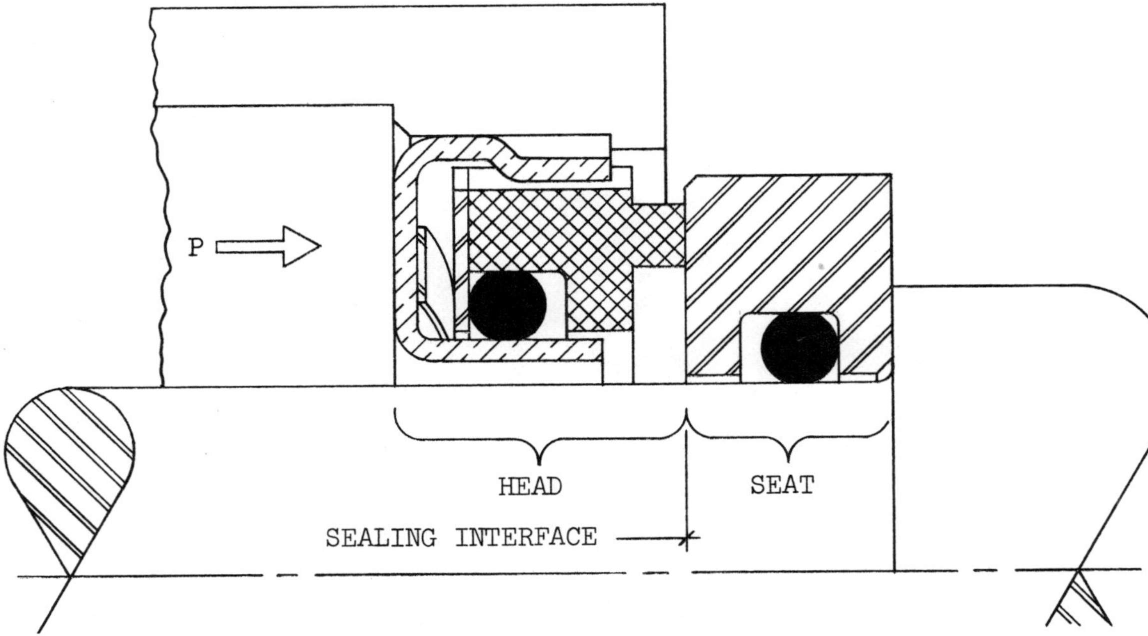

Fig. 2-4b Internal Pressure, Underbalanced

1. *Underbalanced Seal*—An underbalanced mechanical face seal is one which, under a given set of operating conditions, the applied hydraulic pressure, or sealed system operating pressure, results in a net hydraulic force which tends to open the seal.
2. *Balanced Seal*—A term applied to a seal constructed so that, under a given set of operating conditions, the net hydraulic force acting to open or close the seal is essentially zero. In this design, the only significant face load acting to close the seal is mechanical in nature; usually a spring device.
3. *Overbalanced Seal*—An overbalanced mechanical face seal is one which, under a given set of operating conditions, the applied hydraulic pressure, or sealed system pressure, results in a net hydraulic force which tends to close the seal.

D. SEALING FACE CONDITIONS

1. Face Flatness

Good face seal performance is dependent upon the flatness of the mating sealing surfaces which, in the majority of applications, are flat to within 2 to 5 helium light bands (see Section VI). This means that the profile of the whole sealing surface, excluding the dubbed edges and random defects in the surface, lie between two parallel planes located 24 to 60 microinches apart.

Sealing faces should properly be specified as being flat within a certain amount per inch of diameter, rather than an arbitrary amount, regardless of face diameter, since the former method more accurately reflects the increasing difficulty, thus increasing cost, of obtaining flatness on larger diameter faces. Secondly, the same amount of flatness per inch of diameter is seldom required on both of the opposing sealing faces, since one is normally softer and more flexible, and will thus wear into and conform to the much more wear-resistant and rigid counterface, which must be held to a greater degree of flatness since the latter acts as the reference or control surface.

The initial flatness required on any given set of sealing faces depends upon the materials of construction, the design of the seal, the conditions under which the seal operates, and the amount of initial and long-term leakage allowed.

The flatness of functioning seal faces may be adversely affected by excessive or localized application of spring forces, hydraulic pressures, thermal expansions and contractions, locking notches, holes and ears which destroy symmetry and cause uneven stresses, by concentrated torque restraints, dimensional changes in confined elastomeric secondary seals, and by rough handling of seal components before and during installation. In some cases, parts will creep out of flatness with time or temperature due to residual stresses in the part. In addition, load-bearing elements, such as bearing raceways, are subject to mechanical and thermal distortion.

The flatness of the sealing faces is normally 100% inspected by the manufacturer, using a monochromatic light source and an optical flat. More information about inspecting mechanical face seals may be found in Section VI.

2. Face Finish

a. *Roughness*—Specifying a roughness range on a sealing surface can be misleading, since the method of measurement and the materials being measured affect the range to be specified and the significance of the measurement. Surface roughness is an arithmetical deviation of the measured profile to a nominal profile and does not include random defects in the surface being measured. Surface roughness measurements have limited accuracy and are basically qualitative rather than quantitative.

The microscopic structure of the sealing surface material has a large influence on the allowable range of roughness. Some materials, such as carbon-graphite and powdered metals, tend to give roughness readings greater than conventional materials when machined in a similar manner. Some carbon-graphite surfaces will give roughness rms readings approximately 10 times those obtained from conventional metal surfaces after lapping, because the porosity of the material itself exceeds the actual surface roughness. Thus, such readings are substantially meaningless. In terms of performance, roughness finish on many carbon-graphite parts is not critical, because of the self-lapping characteristics of the material.

A common mistake is to equate reflectivity with smoothness on a seal surface. A surface with a highly polished and reflective finish may actually be rougher than another with an unpolished or dull matte finish. In the case of seal faces, the reflectivity of the surface obtained depends initially upon the type of lapping process used, but may be altered by subsequent machine or manual polishing operations to produce either a high polish or a semipolished finish.

b. *Reflectivity*—Polishing of sealing surfaces is often specified to enable inspection of these surfaces with a monochromatic light source and an optical flat, to mask the scratches inherent in the lapping process, to improve the cosmetic attractiveness of the part, and in some cases to reduce the surface roughness of the part. Since polishing tends to reduce the roughness of the sealing surface, it must be

Flatness Measuring Instrument with Meter and Recorder—Provides precise measurement and pictorial printout of the amount and type of out-of-flatness of an individual seal face with an accuracy of two millionths of an inch.

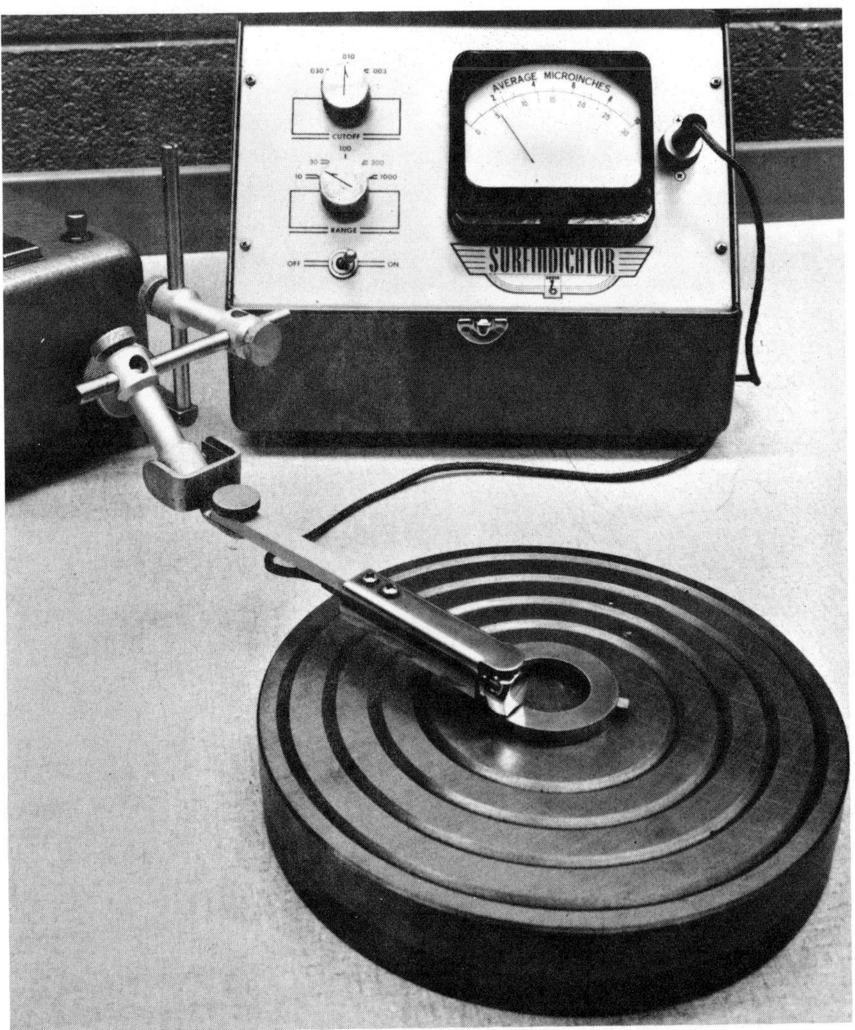

Surfindicator * – Tracer device for measurement of surface roughness in the range of 1 to 1,000 microinches, with direct readout.

* Trademark.

specified with discretion. Too smooth a surface on certain materials—ceramics, for example—often degrades the initial performance of seals operating under mixed or boundary lubrication conditions, such as may be experienced when sealing water-based fluids.

c. *Scratches*—Specifying a sealing surface as simply "free of scratches" should be avoided. Lapping, the method most commonly used to produce a flat sealing surface on hard materials, is an abrasive machining process that inherently imparts literally millions of multi-directional scratches to the surface being machined. Since many surfaces may be used in the as-lapped condition, these scratches themselves are often not harmful but, in many cases, are beneficial to the performance of a seal.

Scratches or burnish lines generally tend to affect leakage only when they are directional, excessive in magnitude and number, and span the width of the sealing interface.

Even if scratches or burnish lines are excessive, the increase in the magnitude of leakage will in most cases be small and short lived, because most scratches will be smeared over or filled in by material transfer, or with residue of the fluid being sealed. These phenomena constitute the self-healing properties of a seal, and often take place during the operating life of a seal when wear debris or abrasive particles temporarily exist in the sealing interface, causing a degree of scratching or burnishing.

The apparent magnitude and direction of scratches and burnish marks are deceptive and are dependent upon the basic reflectivity of the material, the overall roughness of the material, the degree of reflectivity the surface has been given during lapping and polishing, and the intensity and incidence of the light rays illuminating the surface in relation to the viewer. A good approximation of the real magnitude of a scratch can only be derived with the careful use of a precision optical measuring device.

d. *Chips*—Chips result from small material breakouts occurring at an edge of a sealing face during normal part manufacturing and handling procedures. The magnitude of a chip is usually defined by the amount it intrudes into the sealing surface of a seal. Specifying a sealing surface edge as "no chips allowed" should be avoided, because almost any edge, if examined closely, will be found to have some small chips in it.

An acceptable specification for chips is that no limit be placed on the number of chips that intrude into the radial sealing surface less than 10% of its radial width; that up to three chips be allowed that intrude into the surface more than 10%, but less than 15%; and that only one chip be allowed, on a given edge, to intrude into the sealing

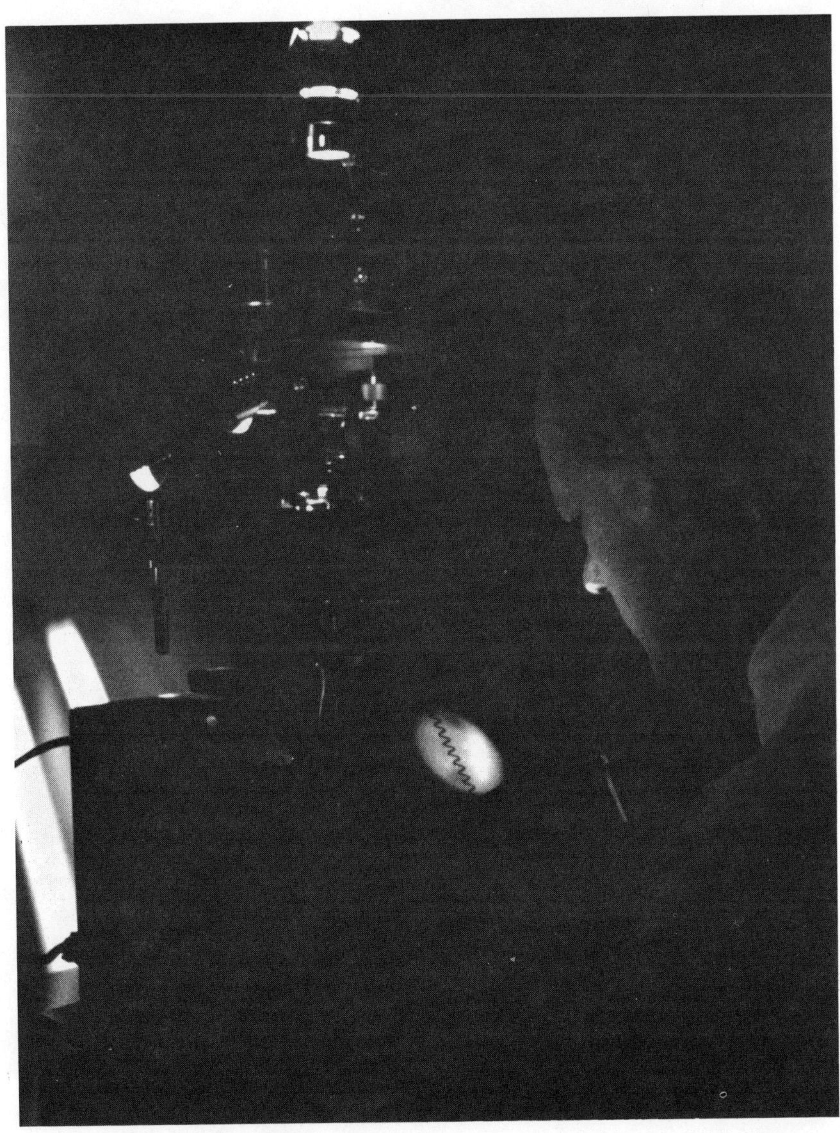

Metallograph—Metallurgical transistorized microscope with magnification range of 5× to 2200×, equipped with filters, screen projection and photo recording facilities. Used for inspection of material microstructures, flaws and inclusions.

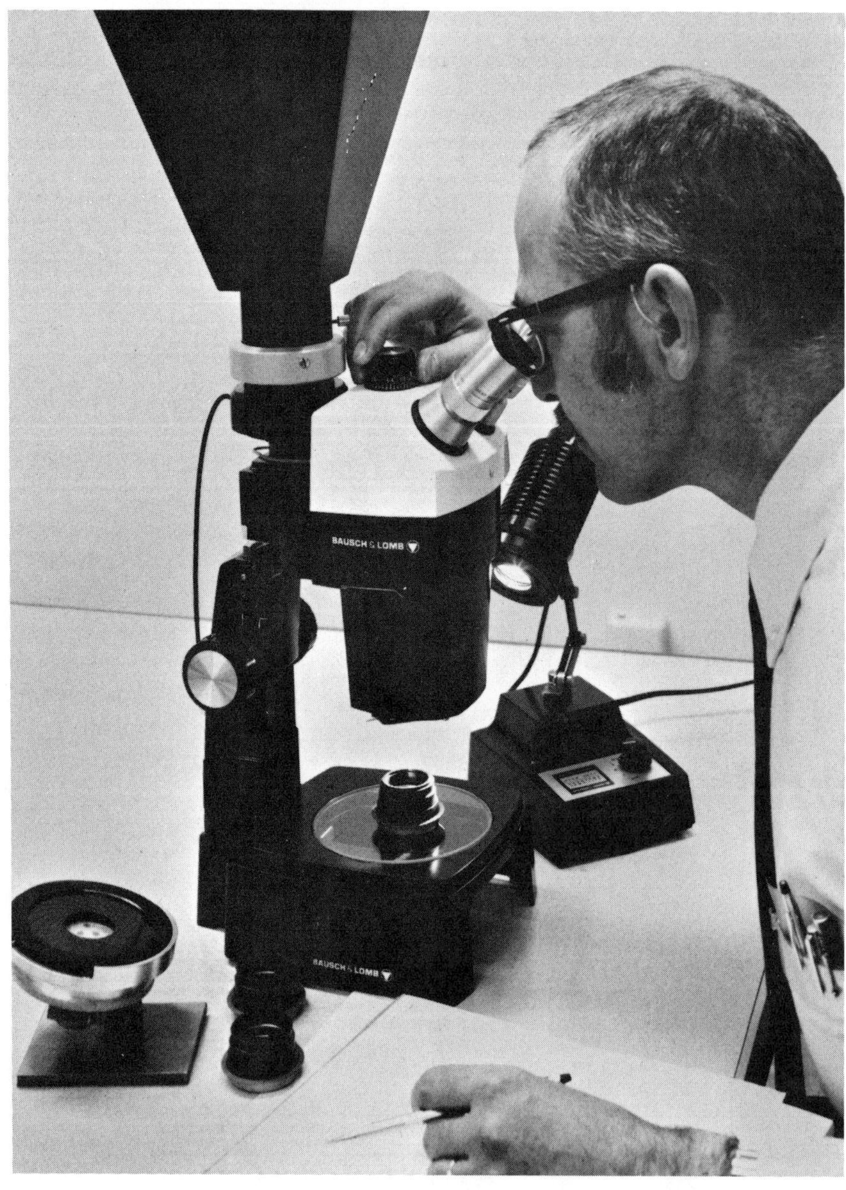

Stereo Zoom Microscope – Magnifies anywhere from 5× to 70×. Equipped with high intensity illuminator, sub-stage lighting and 4″ x 5″ Polaroid camera back. Used for inspecting and recording seal face defects, wear patterns, metal failures and surface conditions.

area more than 15%, but less than 20%. This means that with a 0.090-inch radial-width seal face, no limit would be placed on the number of chips that intrude 0.009 inch or less; up to three could intrude between 0.010 and 0.013 inch; and that only one chip could intrude between 0.014 and 0.018 inch per edge. Except when coarse abrasives, refrigerants, or gases are to be sealed, using a tighter specification than this on seal faces usually results in increased costs for insignificant improvements in performance.

E. SEAL PRELOADING

1. Amount Required

The preloading of the seal, in a balanced or overbalanced design, usually need only be sufficient to overcome secondary seal friction or hysteresis, the friction of any positive drive devices, and the dynamic inertia of the moveable seal components. The preloading should be sufficient to overcome these factors at the greatest axial ranges at which the seal is expected to operate, taking into account tolerances and projected wear rates, as well as allowing for preloading tolerances. Computations of secondary seal hysteresis should, in the case of elastomeric secondary seals, include the effects of operating temperature extremes and heat aging. Moreover, if the application involves hydraulic pressure reversals, or periods of vacuum operation, these conditions must also be taken into account. Unnecessarily high preloading will tend to shorten the useful life of the seal, especially due to excessive friction during start-ups and coast-downs and any periods of dry operation.

2. Methods of Preloading

a. *Single Helical Compression Spring*—Probably the most commonly used method of preloading mechanical face seals, the single helical compression spring offers the advantages of low costs and resistance to solids, abrasives, and corrosive conditions. Care must be exercised in designing the single helical compression spring, and the seal in which it is used, to insure uniform loading around the annular area of the sealing faces. Preloading by the use of a single helical compression spring is limited to seals having sufficient space and, in the case of rotating-head type seals, where rotational speeds do not cause the coils of the spring to unwind and expand excessively due to centrifugal force, resulting in malfunctioning of the seal.

b. *Multiple Helical Compression Springs*—The use of multiple helical compression springs provides good even preloading around the annular area of the sealing faces, and allows a more compact seal de-

Instron Universal Testing Instrument (10,000 lb. capacity)—Conventional stress-strain tests in tension, compression, or flexure can be made with great accuracy and detail. A number of other characteristics can also be measured and recorded, such as cyclic hysteresis of mechanical face seal assemblies, stress relaxation and recovery of elastomeric components, and press-fit forces of seal housings.

sign than possible with a single helical compression spring. In addition, multiple springs are often more advantageously employed in rotating-head type seals for high speed applications than is the case with a single helical compression spring. Designs making use of multiple helical compression springs are often more expensive and sensitive to solids, abrasive conditions, and corrosives, than designs with only a single helical compression spring.

c. *Single Wave Washer Spring*—The use of this type of spring provides excellent evenness of preloading around the annular area of the sealing faces, and allows very compact seal design. Disadvantages lie in the tempering requirements which limit material choices to those which are often unsuitable for severe corrosive applications. In addition, wave washer springs, as a class, have a relatively high load rate which results in a greater change in loading for a given deflection.

d. *Single Belleville Washer Spring*—In general, this type has the same advantages and disadvantages as the wave washer spring, except the nonlinear load-deflection characteristic of this type of spring can be used to advantage if the design of the seal permits its use.

e. *Stacked Wave Washer Springs*—Wave washer springs can be assembled in parallel (Fig. 2-5) increasing the load in proportion to the number of springs. However, they cannot be used in series (Fig. 2-6) in order to increase deflection, without using some means of keeping the waves of adjacent springs aligned, unless the springs are separated by flat spacer washers. If the space for flat spacer washers cannot be allowed, adjacent springs must be aligned by riveting, by spot welding or by the use of keyways, but these methods can cause undesirable stress concentrations within the wave washer springs.

f. *Stacked Belleville Washer Springs*—Belleville washer springs may be stacked in series (Fig. 2-7) often without the necessity of using flat spacer washers. Like wave washer springs they can also be assembled in parallel (Fig. 2-8) increasing the load in proportion to the number of springs.

g. *Elastomeric Preloading*—This type of preloading is sometimes used in seals operating under low speeds and pressures, such as heavy-duty "off-the-road" service seals, and low-cost appliance seals. The most popular form of these elastomeric "springs" is that of a conical disk or belleville washer, such designs being patented in the late nineteen thirties and early forties. As well as serving to preload the seal, these elastomeric members also usually fulfil the secondary sealing and anti-rotation requirements of the seal design. (See Figs. 1-6 and 2-30.) Other designs which take advantage of the resilient proper-

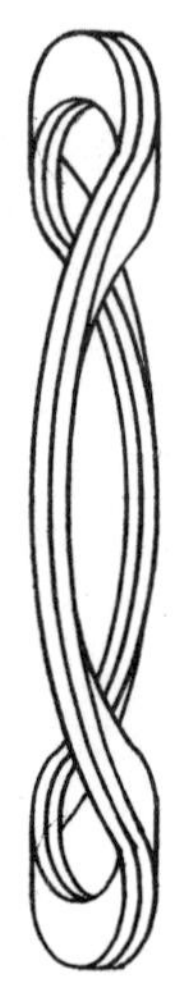

Fig. 2-5
Wave Springs in Parallel

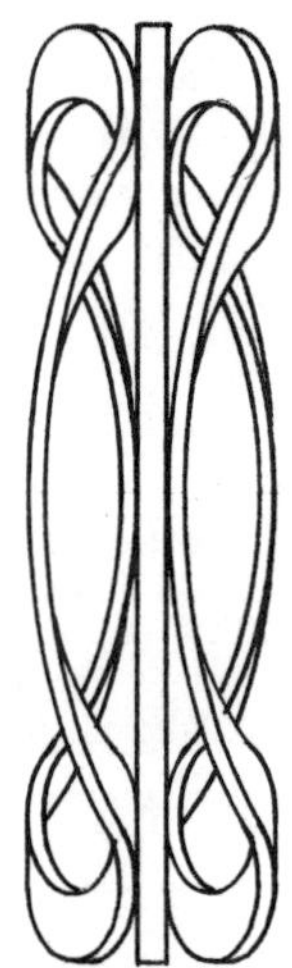

Fig. 2-6
Wave Springs in Series

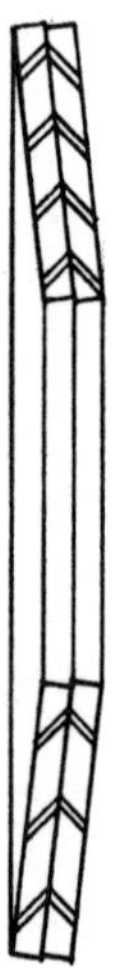

Fig. 2-7
Belleville Springs in Parallel

Fig. 2-8
Belleville Springs in Series

ties of elastomers for preloading make use of diaphragms, O-rings, grommets, and other shapes. (See Fig. 1-9)

TABLE I
Maximum Recommended Service Temperatures of Selected Materials

Seal Face Materials and Platings

Material	Temp (deg F)
Filled Teflon	500
Bronze	500
Low Alloy Gray Irons	650
Malleable Iron	700
Ductile Iron	700
Ni-Resist	800
410 Stainless Steel	900
17-4 PH Stainless Steel	900
Tool Steel R_c62-65	900
S-Monel	950
Ductile Ni-Resist	1000
High Temperature Carbon	1200
Stellite #31	1200
Inconel X	1200

Plating	Temp (deg F)
Chromium	500
Silver	600
Tin	700
Cadmium-nickel	1000
LW1	1000
LC-1A	1600
LA-2	1600

Coil Spring Materials

Material	Temp (deg F)
Phosphor-bronze ASTM B159	200
Silicon-bronze ASTM B99	200
Ni Span C 902	200
Music ASTM A228	250
Hard-drawn ASTM A227	250
Oil-tempered ASTM A229	300
Valve-spring ASTM A230	300
Beryllium-copper ASTM B197	400
Chrome-vanadium AISI 6150	425
Silicone-manganese AISI 9260	450
Chrome-silicone AISI 9254	475
Martensitic AISI 410	500
Martensitic AISI 420	500
Austenitic AISI 301	600
Austenitic AISI 302	600
17-7 PH Stainless Steel	700
Inconel 600	700
Nickel-chrome A286	950
Inconel 718	1200
Inconel X-750	1300
L-605	1400
S-816	1400
Rene 41	1400

TABLE I (Cont'd)

Flat Spring Materials

Material	Temp (deg F)
Ni-Span C 902	200
Phosphor-bronze ASTM B103	200
High-carbon AISI 1050	200
High-carbon AISI 1065	200
High-carbon AISI 1075	250
High-carbon AISI 1095	250
Beryllium-copper ASTM B194	400
Austenitic AISI 301	600
Austenitic AISI 302	600
17-7 PH Stainless Steel	700
Inconel 600	700
Beryleo Nickel	700
Titanium 6-6-2	750
Sandvik 11R51	800
Duranickel 301	800
Permanickel	800
Elgiloy	900
Havar	900
Inconel 718	1200
Inconel X-750	1300
Rene 41	1400

Formed Metal Bellows Materials

Material	Temp (deg F)
Brass CDA 240	300
Phosphor-bronze CDA 510	300
Beryllium-copper CDA 172	350
Monel 404	450
Unstabilized 300 Series Stainless Steel	500
Ni-Span C 902	750
Stabilized 300 Series Stainless Steel	800
Inconel 600	800
Inconel X-750	1500

Welded Metal Bellows Materials

Material	Temp (deg F)
Ni-Span C	500
AM-350 Stainless Steel	800
410 Stainless Steel	800
Commercially Pure Titanium	800
Stabilized 300 Series Stainless Steel	1200
Inconel X-750	1500
Inconel 625	1500
Hastelloy-C	1800
Rene 41	1800

The advantage of preloading a seal by use of an elastomeric component lies in inherent cost savings, since some form of elastomeric member is usually required anyway to provide for a flexible mounting or for a secondary sealing function. The disadvantages of using this type of preloading are that elastomeric materials do not have constant physical properties, and loads vary widely during the service life of the seal due to stress relaxation, permanent set, low temperature crystallization, heat aging, high thermal expansions and contractions, swelling and shrinking in various fluids, and deterioration of elastomers due to chemical and ozone attack.

h. *Magnetic Preloading*—Using magnetic force to preload mechanical face seals usually results in a compact seal design of few parts. Unfortunately, such designs usually feature short axial seal faces, and present difficulties in manufacturing, packaging, shipping, storage and handling. In addition, in most systems being sealed there are usually metallic wear particles and foreign matter present which are attracted to magnetically preloaded seals. Material selection is also limited in seal designs which rely on magnetic force for preloading.

i. *Formed Metal Bellows*—Formed metal bellows can be used to preload, and to provide for the secondary sealing and antirotation requirements of mechanical face seals. Designs incorporating formed metal bellows are usually more expensive, and require more space, than seals incorporating more conventional methods of preloading; their use therefore is usually restricted to applications requiring the temperature range capabilities of a metal secondary seal.

j. *Welded Metal Bellows*—Although even more expensive than formed metal bellows, welded metal bellows are usually used in mechanical face seals operating at extreme temperatures, because of their superior reliability. Welded metal bellows can be made more compact than formed bellows; and they are available in a wide variety of materials suitable for extreme environmental temperatures. In addition to providing for seal preload, welded metal bellows also satisfy the secondary sealing and anti-rotation requirements of mechanical face seals.

k. *Garter Spring*—Mechanical face seals have been built which employ the use of a garter spring acting against a conical surface to provide the necessary axial preloading. Such designs, however, require a rather large amount of space and are quite sensitive to solids and abrasives.

F. POSITIVE DRIVE METHODS

Many different methods are employed in mechanical face seals to rotate components, or to prevent parts from rotating in operation. Some of these methods allow axial movement of the component while preventing rotational movement with respect to another adjacent component. See Figures 2-9 through 2-34.

Positive Drive Methods

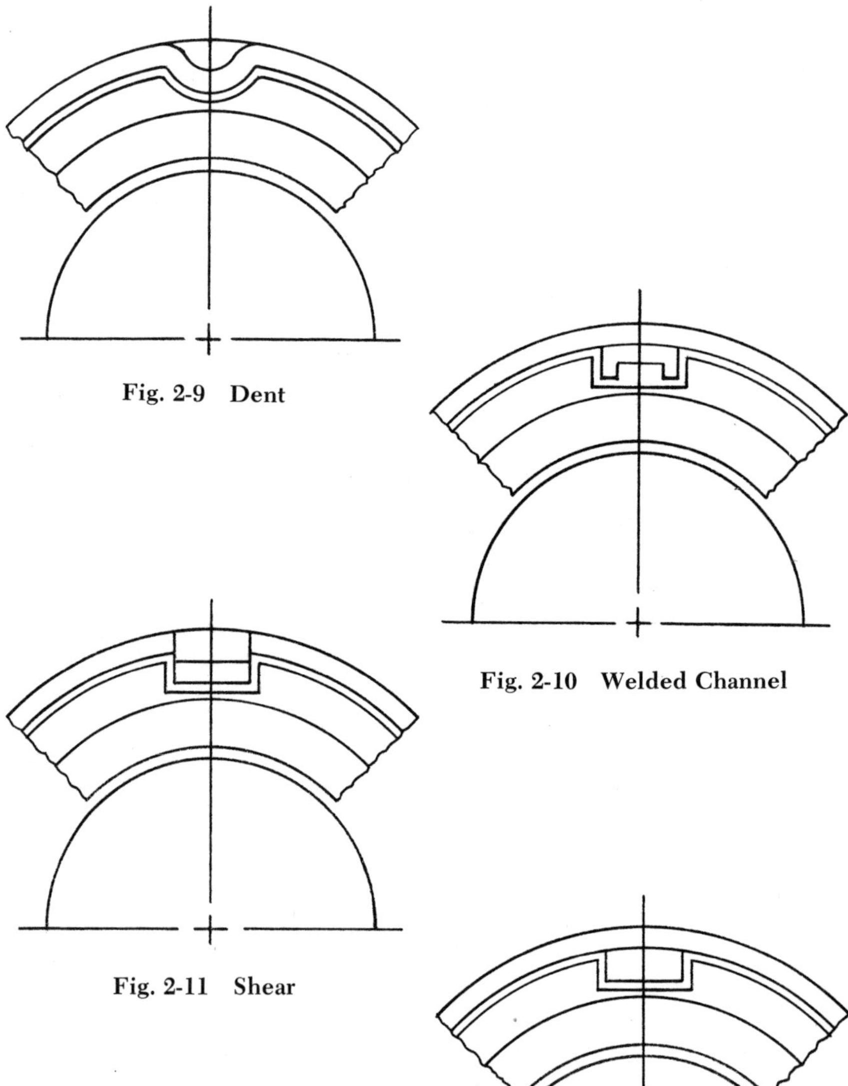

Fig. 2-9 Dent

Fig. 2-10 Welded Channel

Fig. 2-11 Shear

Fig. 2-12 Welded Lug

Positive Drive Methods (Cont'd)

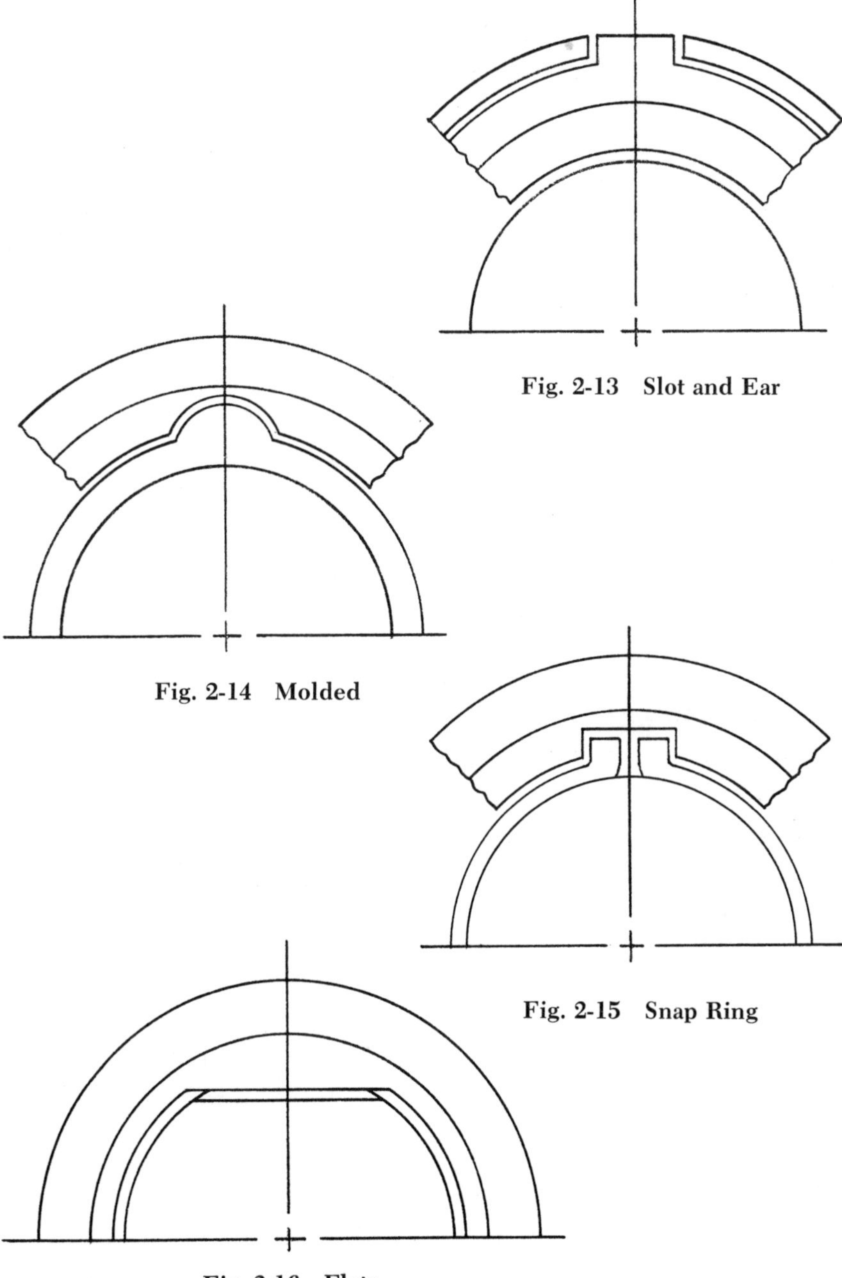

Fig. 2-13 Slot and Ear

Fig. 2-14 Molded

Fig. 2-15 Snap Ring

Fig. 2-16 Flats

Positive Drive Methods (Cont'd)

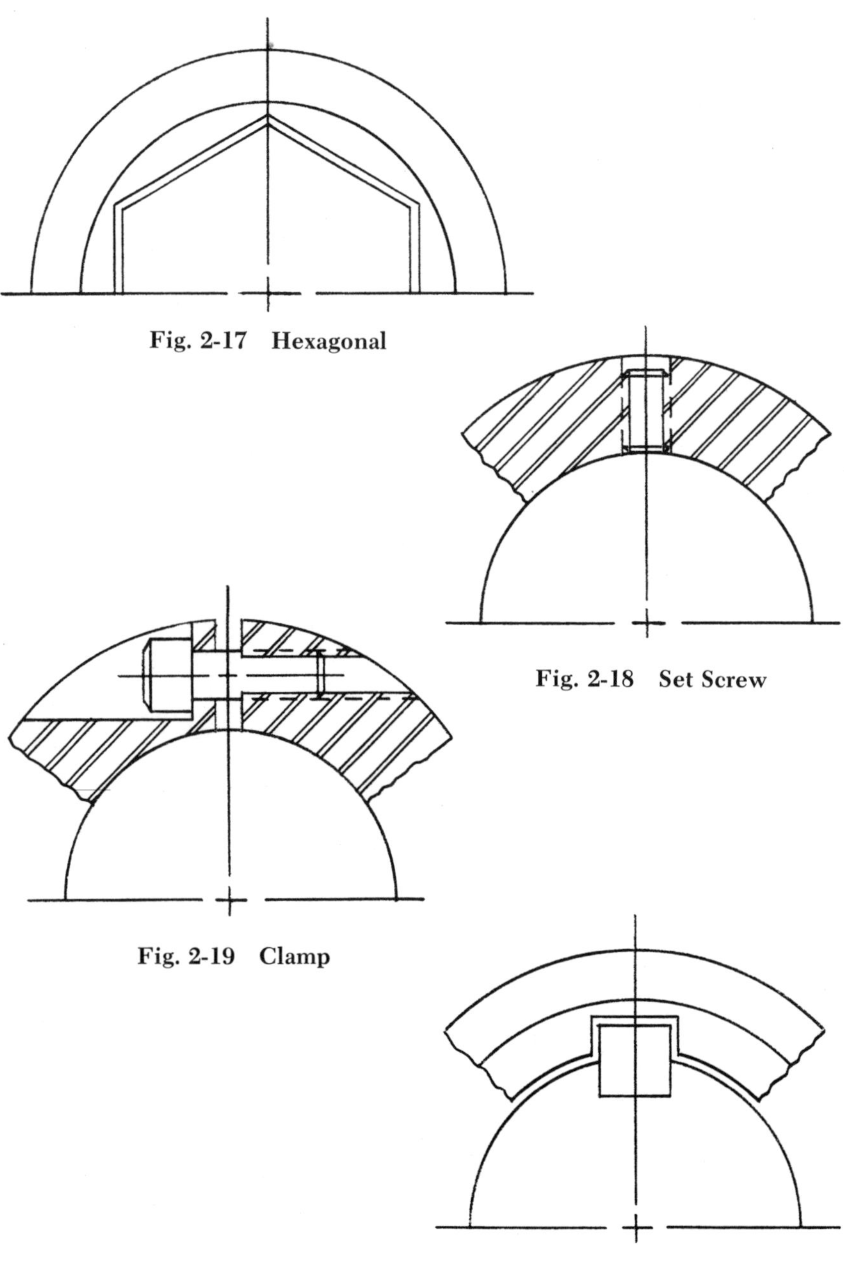

Fig. 2-17 Hexagonal

Fig. 2-18 Set Screw

Fig. 2-19 Clamp

Fig. 2-20 Key

Positive Drive Methods (Cont'd)

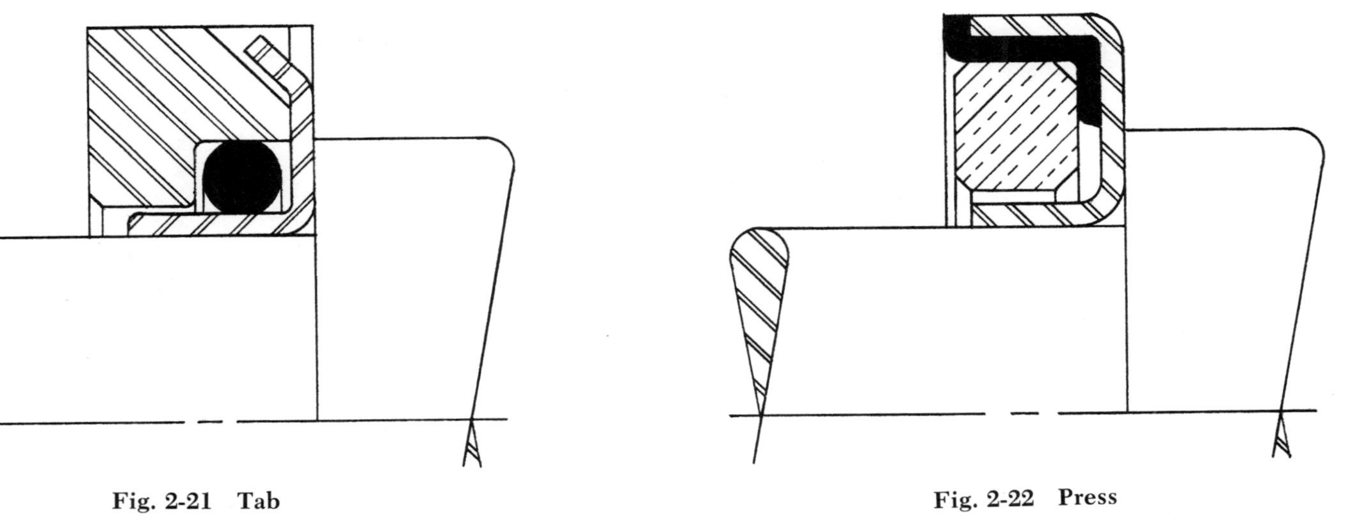

Fig. 2-21 Tab

Fig. 2-22 Press

Positive Drive Methods (Cont'd)

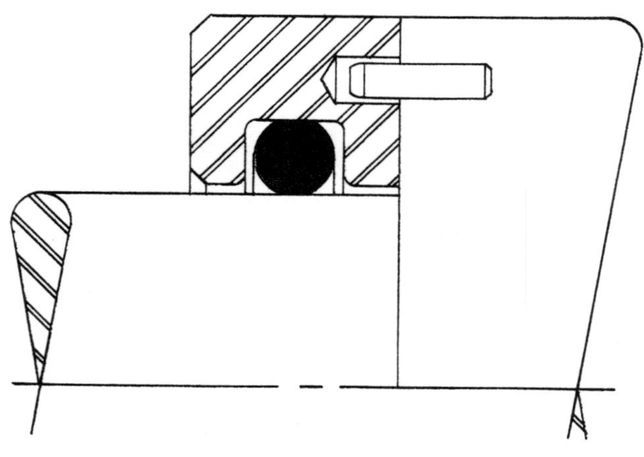

Fig. 2-23 Pin, Dowel, Roll Pin

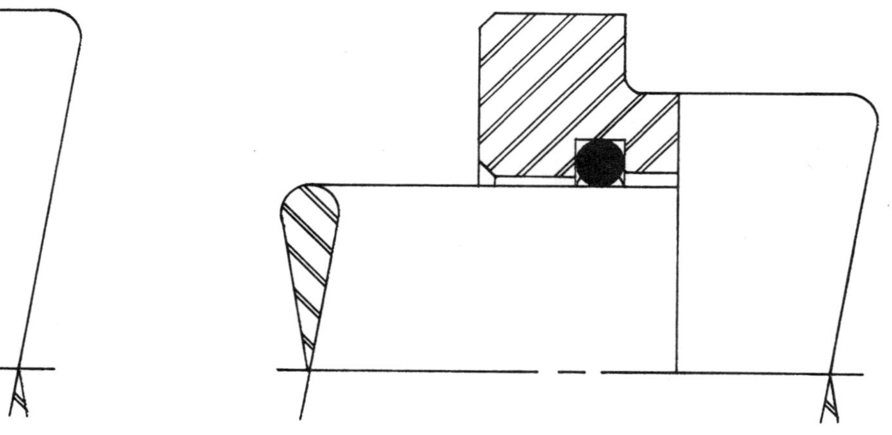

Fig. 2-24 Squeezed Elastomer

Positive Drive Methods (Cont'd)

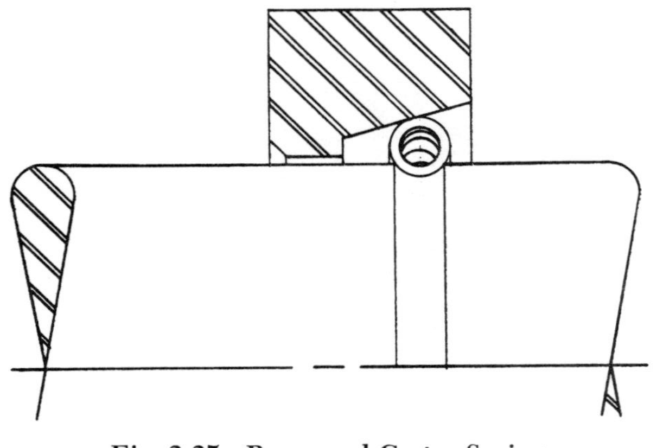

Fig. 2-25 Ramp and Garter Spring

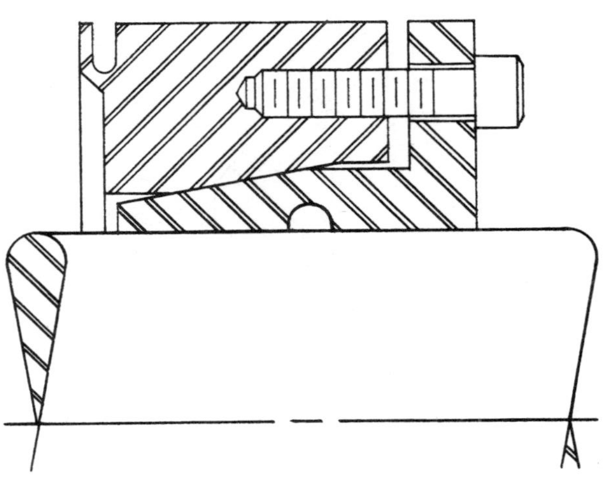

Fig. 2-26 Mechanical Wedge

Positive Drive Methods (Cont'd)

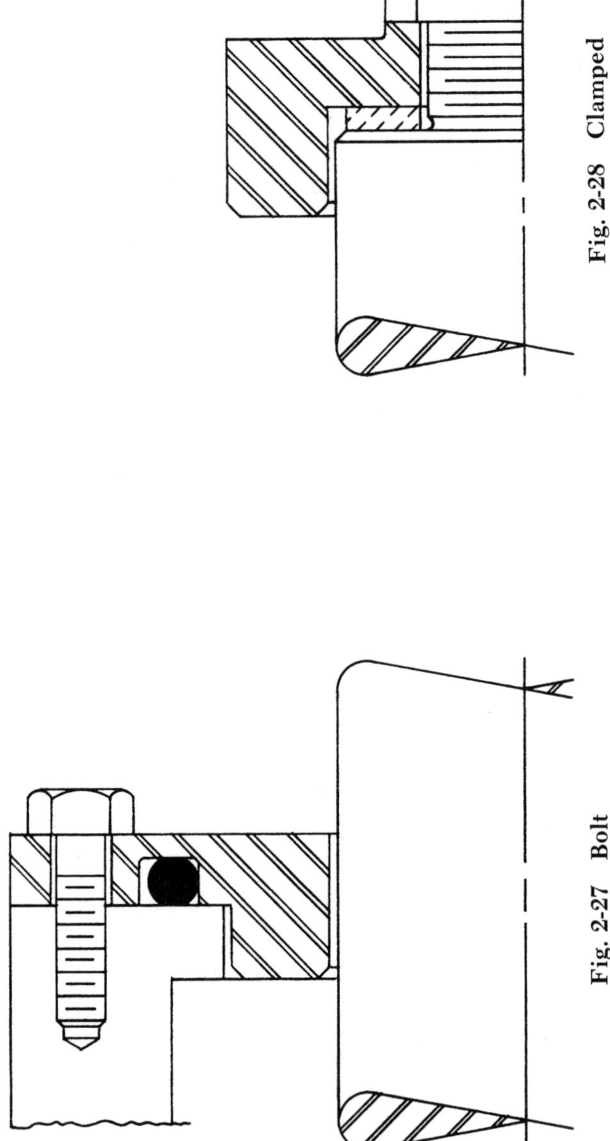

Fig. 2-27 Bolt

Fig. 2-28 Clamped

Positive Drive Methods (Cont'd)

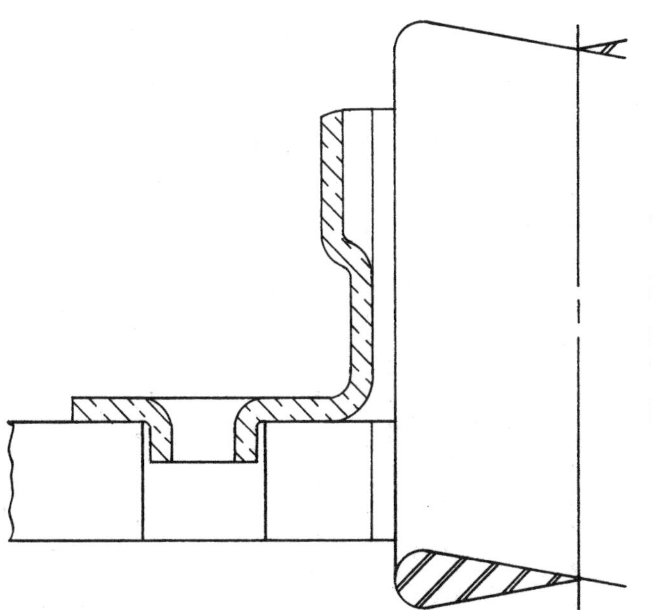

Fig. 2-29 Pierced

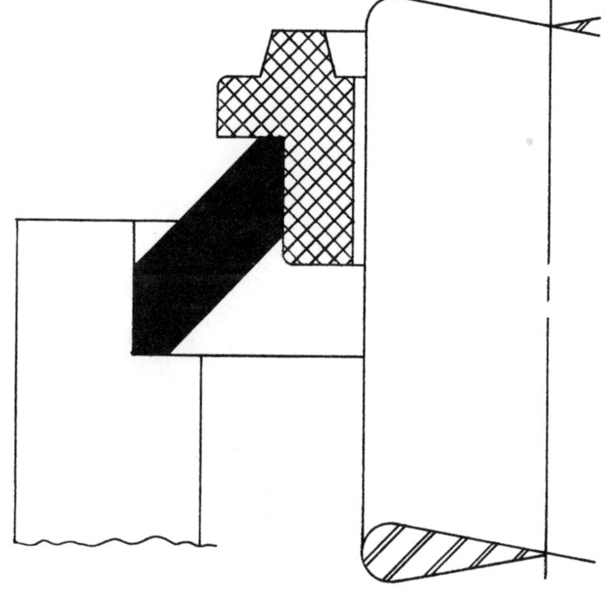

Fig. 2-30 Wedged Elastomer

Positive Drive Methods (Cont'd)

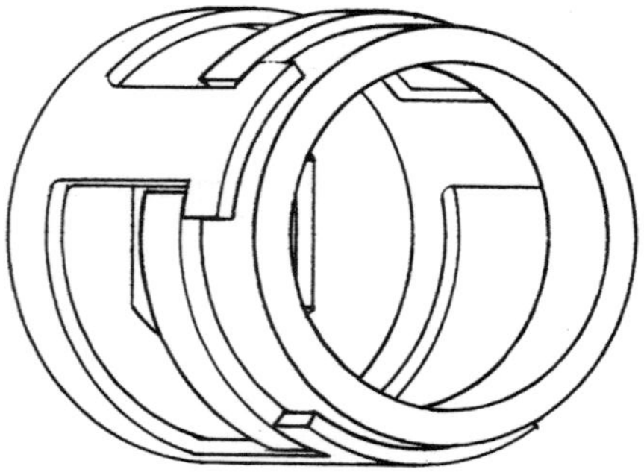

Fig. 2-31 Finger

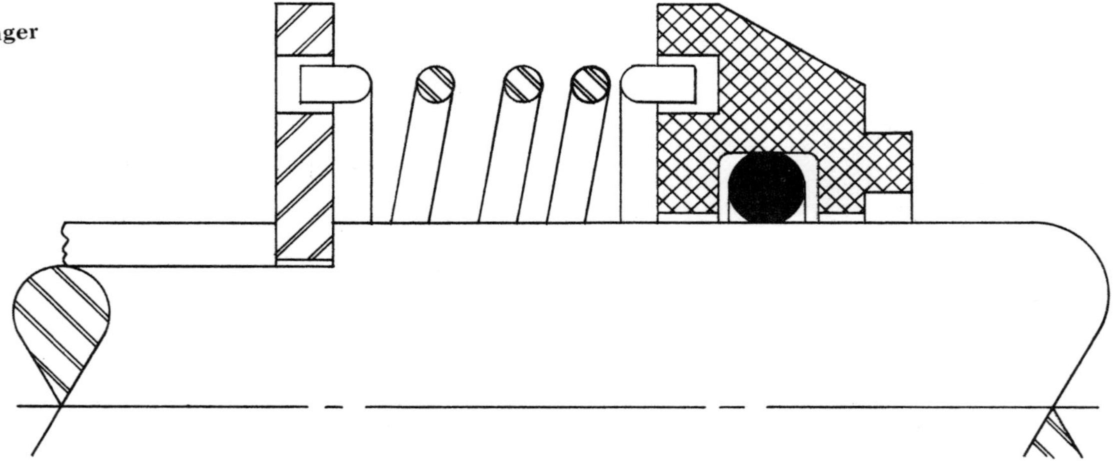

Fig. 2-32 Spring

Positive Drive Methods (Cont'd)

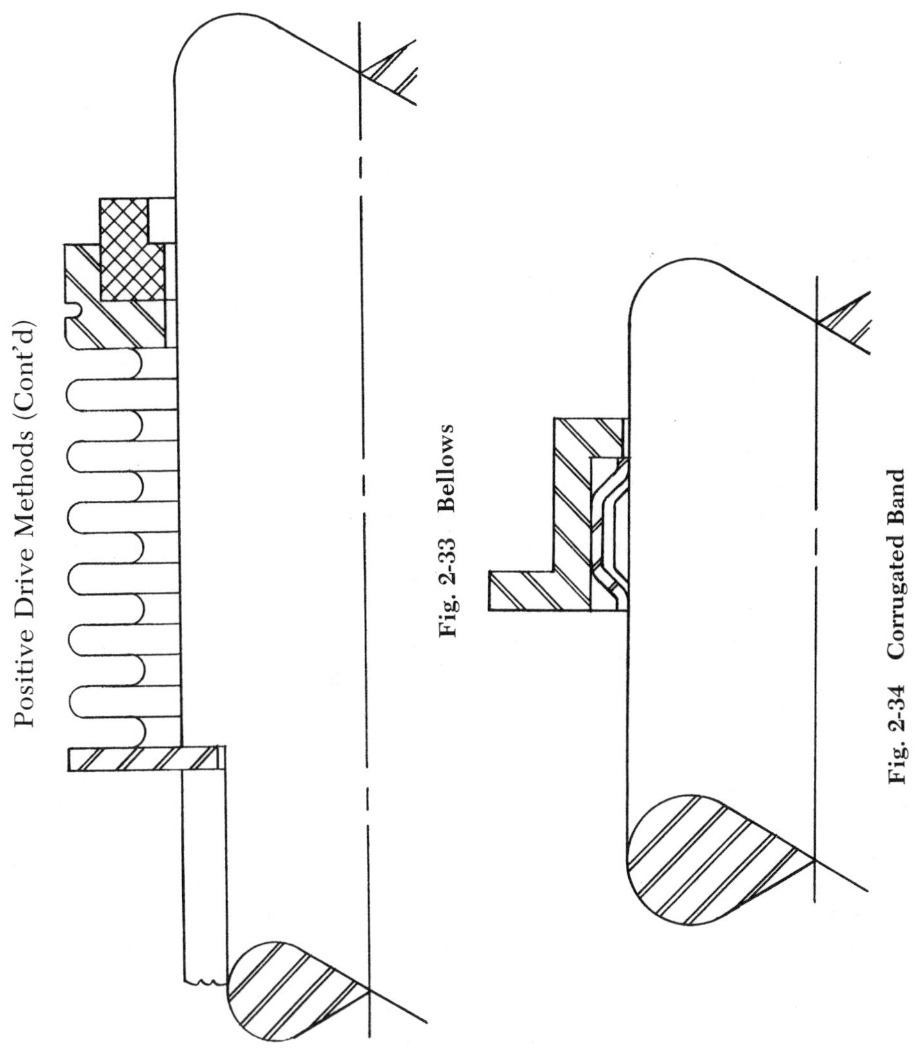

Fig. 2-33 Bellows

Fig. 2-34 Corrugated Band

G. USE OF O-RINGS

O-rings are often used as secondary seals in mechanical face seal designs, or to assure a pressure-tight installation of a seal component or assembly in a bore or around a shaft; they are also sometimes used to prevent relative rotation between a seal component, or assembly, and the adjacent shaft or bore. Among the reasons for the extensive use of O-rings are their availability in many sizes and materials, low cost, ease of installation, adaptability to a limited space, simplicity of design, and overall effectiveness.

1. Operation

Normally an O-ring is installed in a groove and effects a seal through mechanical squeeze, augmented by pressure. Primary seal is made at installation when the O-ring is squeezed between adjacent surfaces, normally in a radial direction. Under fluid pressure, the O-ring is forced axially against the opposing wall of the groove in which it is installed, and intensifies the radial sealing action.

2. Application

O-rings seal against pressure originating from either side and may be used in a dynamic (moveable) or static (fixed) relationship with adjacent surfaces when sealing gas or liquid in a wide variety of applications. O-rings are commonly used to seal vacuum, or pressures to 1,500 psi without anti-extrusion devices. At higher pressures, exceeding 1,500 psi, anti-extrusion devices made of metal, plastic or leather are usually required.

3. Selection

a. *Size*—O-rings are available in various standard sizes. The sizes most commonly used with mechanical face seals, and their respective dimensions, may be found in Table II.

b. *Material*—In any particular application, the compound of which the O-ring is made is usually based on the polymer having the optimum combination of characteristics matching the requirements of that application. To this base material, from a few to almost two dozen ingredients (such as reinforcing agents, curatives, plasticizers, and lubricants) may be added during compounding, which further tailors the compound towards the end requirements of the application. An O-ring compound must meet many requirements in order to function effectively; these requirements include:

TABLE II

O-Ring Sizes Commonly Used with Mechanical Face Seals

ARP 568 Dash No.	Actual ID Max.	Actual ID Min.	Cross-section Max.	Cross-section Min.	(Ref) OD Max.	(Ref) OD Min.
–010	.244	.234	.073	.067	.390	.368
–011	.306	.296	.073	.067	.452	.430
–012	.369	.359	.073	.067	.515	.493
–013	.431	.421	.073	.067	.577	.555
–014	.494	.484	.073	.067	.640	.618
–015	.556	.546	.073	.067	.702	.680
–016	.619	.609	.073	.067	.765	.743
–017	.681	.671	.073	.067	.827	.805
–018	.744	.734	.073	.067	.890	.868
–019	.806	.796	.073	.067	.952	.930
–020	.870	.858	.073	.067	1.016	.992
–021	.932	.920	.073	.067	1.078	1.054
–022	.995	.983	.073	.067	1.141	1.117
–023	1.057	1.045	.073	.067	1.203	1.179
–024	1.120	1.108	.073	.067	1.266	1.242
–025	1.182	1.170	.073	.067	1.328	1.304
–026	1.245	1.233	.073	.067	1.391	1.367
–027	1.307	1.295	.073	.067	1.453	1.429
–028	1.370	1.358	.073	.067	1.516	1.492
–029	1.499	1.479	.073	.067	1.645	1.613
–030	1.624	1.604	.073	.067	1.770	1.738
–031	1.749	1.729	.073	.067	1.895	1.863
–032	1.874	1.854	.073	.067	2.020	1.988
–033	1.999	1.979	.073	.067	2.145	2.113
–034	2.124	2.104	.073	.067	2.270	2.238
–035	2.249	2.229	.073	.067	2.395	2.363
–036	2.374	2.354	.073	.067	2.520	2.488
–037	2.499	2.479	.073	.067	2.645	2.613
–038	2.624	2.604	.073	.067	2.770	2.738
–039	2.754	2.724	.073	.067	2.900	2.858
–040	2.879	2.849	.073	.067	3.025	2.983
–041	3.004	2.974	.073	.067	3.150	3.108
–042	3.254	3.224	.073	.067	3.400	3.358
–043	3.504	3.474	.073	.067	3.650	3.608
–044	3.754	3.724	.073	.067	3.900	3.858
–045	4.004	3.974	.073	.067	4.150	4.108
–046	4.254	4.224	.073	.067	4.400	4.358
–047	4.504	4.474	.073	.067	4.650	4.608

(Cont'd)

TABLE II (Cont'd)

ARP 568 Dash No.	Actual ID Max.	Actual ID Min.	Cross-section Max.	Cross-section Min.	(Ref) OD Max.	(Ref) OD Min.
–048	4.754	4.724	.073	.067	4.900	4.858
–049	5.012	4.966	.073	.067	5.158	5.100
–110	.367	.357	.106	.100	.579	.557
–111	.429	.419	.106	.100	.641	.619
–112	.492	.482	.106	.100	.704	.682
–113	.554	.544	.106	.100	.766	.744
–114	.617	.607	.106	.100	.829	.807
–115	.679	.669	.106	.100	.891	.869
–116	.742	.732	.106	.100	.954	.932
–117	.805	.793	.106	.100	1.017	.993
–118	.868	.856	.106	.100	1.080	1.056
–119	.930	.918	.106	.100	1.142	1.118
–120	.993	.981	.106	.100	1.205	1.181
–121	1.055	1.043	.106	.100	1.267	1.243
–122	1.118	1.106	.106	.100	1.330	1.306
–123	1.180	1.168	.106	.100	1.392	1.368
–124	1.243	1.231	.106	.100	1.455	1.431
–125	1.305	1.293	.106	.100	1.517	1.493
–126	1.368	1.356	.106	.100	1.580	1.556
–127	1.430	1.418	.106	.100	1.642	1.618
–128	1.493	1.481	.106	.100	1.705	1.681
–129	1.559	1.539	.106	.100	1.771	1.739
–130	1.622	1.602	.106	.100	1.834	1.802
–131	1.684	1.664	.106	.100	1.896	1.864
–132	1.747	1.727	.106	.100	1.959	1.927
–133	1.809	1.789	.106	.100	2.021	1.989
–134	1.872	1.852	.106	.100	2.084	2.052
–135	1.935	1.915	.106	.100	2.147	2.115
–136	1.997	1.977	.106	.100	2.209	2.177
–137	2.060	2.040	.106	.100	2.272	2.240
–138	2.122	2.102	.106	.100	2.340	2.302
–139	2.185	2.165	.106	.100	2.397	2.365
–140	2.247	2.227	.106	.100	2.459	2.427
–141	2.310	2.290	.106	.100	2.522	2.490
–142	2.372	2.352	.106	.100	2.584	2.552
–143	2.435	2.415	.106	.100	2.647	2.615
–144	2.497	2.477	.106	.100	2.709	2.677
–145	2.560	2.540	.106	.100	2.772	2.740

TABLE II (Cont'd)

ARP 568 Dash No.	Actual ID Max.	Actual ID Min.	Cross-section Max.	Cross-section Min.	(Ref) OD Max.	(Ref) OD Min.
−146	2.622	2.602	.106	.100	2.834	2.802
−147	2.690	2.660	.106	.100	2.902	2.860
−148	2.752	2.722	.106	.100	2.964	2.922
−149	2.815	2.785	.106	.100	3.027	2.985
−150	2.877	2.847	.106	.100	3.089	3.047
−151	3.002	2.972	.106	.100	3.214	3.172
−152	3.252	3.222	.106	.100	3.464	3.422
−153	3.502	3.472	.106	.100	3.714	3.672
−154	3.752	3.722	.106	.100	3.964	3.922
−155	4.002	3.972	.106	.100	4.214	4.172
−156	4.252	4.222	.106	.100	4.464	4.422
−157	4.502	4.472	.106	.100	4.714	4.672
−158	4.752	4.722	.106	.100	4.964	4.922
−159	5.002	4.972	.106	.100	5.214	5.172
−210	.740	.728	.143	.135	1.026	.998
−211	.802	.790	.143	.135	1.088	1.060
−212	.865	.853	.143	.135	1.151	1.123
−213	.927	.915	.143	.135	1.213	1.185
−214	.990	.978	.143	.135	1.276	1.248
−215	1.052	1.040	.143	.135	1.338	1.310
−216	1.115	1.103	.143	.135	1.401	1.373
−217	1.177	1.165	.143	.135	1.463	1.435
−218	1.240	1.228	.143	.135	1.526	1.498
−219	1.302	1.290	.143	.135	1.588	1.560
−220	1.365	1.353	.143	.135	1.651	1.623
−221	1.427	1.415	.143	.135	1.713	1.685
−222	1.490	1.478	.143	.135	1.776	1.748
−223	1.619	1.599	.143	.135	1.905	1.869
−224	1.744	1.724	.143	.135	2.030	1.994
−225	1.869	1.849	.143	.135	2.155	2.119
−226	1.994	1.974	.143	.135	2.280	2.244
−227	2.119	2.099	.143	.135	2.405	2.369
−228	2.244	2.224	.143	.135	2.530	2.494
−229	2.369	2.349	.143	.135	2.655	2.619
−230	2.494	2.474	.143	.135	2.780	2.744
−231	2.619	2.599	.143	.135	2.905	2.869
−232	2.749	2.719	.143	.135	3.035	2.989
−233	2.874	2.844	.143	.135	3.160	3.114

(Cont'd)

TABLE II (Cont'd)

ARP 568 Dash No.	Actual ID Max./Min.	Cross-section Max./Min.	(Ref) OD Max./Min.	ARP 568 Dash No.	Actual ID Max./Min.	Cross-section Max./Min.	(Ref) OD Max./Min.
−234	2.999/2.969	.143/.135	3.285/3.239	−326	1.610/1.590	.215/.205	2.040/2.000
−235	3.124/3.094	.143/.135	3.410/3.364	−327	1.735/1.715	.215/.205	2.165/2.125
−236	3.249/3.219	.143/.135	3.535/3.489	−328	1.860/1.840	.215/.205	2.290/2.250
−237	3.374/3.344	.143/.135	3.660/3.614	−329	1.985/1.965	.215/.205	2.415/2.375
−238	3.499/3.469	.143/.135	3.785/3.739	−330	2.110/2.090	.215/.205	2.540/2.500
−239	3.624/3.594	.143/.135	3.910/3.864	−331	2.235/2.215	.215/.205	2.665/2.625
−240	3.749/3.719	.143/.135	4.035/3.989	−332	2.360/2.340	.215/.205	2.790/2.750
−241	3.874/3.844	.143/.135	4.160/4.114	−333	2.485/2.465	.215/.205	2.915/2.875
−242	3.999/3.969	.143/.135	4.285/4.239	−334	2.610/2.590	.215/.205	3.040/3.000
−243	4.124/4.094	.143/.135	4.410/4.364	−335	2.740/2.710	.215/.205	3.170/3.120
−244	4.249/4.219	.143/.135	4.535/4.489	−336	2.865/2.835	.215/.205	3.295/3.245
−245	4.374/4.344	.143/.135	4.660/4.614	−337	2.990/2.960	.215/.205	3.420/3.370
−246	4.499/4.469	.143/.135	4.785/4.739	−338	3.115/3.085	.215/.205	3.545/3.495
−247	4.624/4.594	.143/.135	4.910/4.864	−339	3.240/3.210	.215/.205	3.670/3.620
−248	4.749/4.719	.143/.135	5.035/4.989	−340	3.365/3.335	.215/.205	3.795/3.745
−249	4.874/4.844	.143/.135	5.160/5.114	−341	3.490/3.460	.215/.205	3.920/3.870
−250	4.999/4.969	.143/.135	5.285/5.239	−342	3.615/3.585	.215/.205	4.045/3.995
−251	5.132/5.086	.143/.135	5.418/5.356	−343	3.740/3.710	.215/.205	4.170/4.120
−325	1.485/1.465	.215/.205	1.915/1.875	−344	3.865/3.835	.215/.205	4.295/4.245

TABLE II (Concluded)

ARP 568 Dash No.	Actual ID Max. / Min.	Cross-section Max. / Min.	(Ref) OD Max. / Min.	ARP 568 Dash No.	Actual ID Max. / Min.	Cross-section Max. / Min.	(Ref) OD Max. / Min.
–345	3.990 / 3.960	.215 / .205	4.420 / 4.370	–350	4.615 / 4.585	.215 / .205	5.045 / 4.995
–346	4.115 / 4.085	.215 / .205	4.545 / 4.495	–351	4.740 / 4.710	.215 / .205	5.170 / 5.120
–347	4.240 / 4.210	.215 / .205	4.670 / 4.620	–352	4.865 / 4.835	.215 / .205	5.295 / 5.245
–348	4.365 / 4.335	.215 / .205	4.795 / 4.745	–353	4.990 / 4.960	.215 / .205	5.420 / 5.370
–349	4.490 / 4.460	.215 / .205	4.920 / 4.870	–354	5.123 / 5.077	.215 / .205	5.553 / 5.487

i. good resistance to compression set,
ii. chemical compatibility with the service fluid,
iii. stability over the service temperature range,
iv. optimum hardness for operating conditions and requirements,
v. low volume swell in the service fluid,
vi. adequate abrasion resistance,
vii. adequate tear resistance,
viii. enough torsional strength to prevent or withstand spiraling,
ix. adequate ozone resistance,
x. proper coefficient of friction,
xi. sufficient elongation,
xii. ability to be molded to close tolerances.

These requirements show how important it can be to select a compound appropriate to the application. Indeed, it is often necessary to use a material that has been custom compounded for the specific end use intended, if optimum performance is to be realized.

4. Cavity Design

A rectangular cross-section groove is recommended for most applications. Generally the groove width should be 135% of the nominal O-ring cross-section. The outside corners of the groove should be broken .005 inch, and the inside corners rounded to .020 inch. The sides of the groove may slope outwardly from the bottom, up to five degrees. The sides and bottom of the groove should generally be smooth to 30 to 50 microinches RMS, with surfaces that slide in relation to the O-ring held to less than 30 microinches RMS. Cross-sectional squeeze of the O-ring when installed should be no less than

.005 inch regardless of size, but no less than 10% squeeze is usually recommended. Maximum cross-sectional squeeze of the O-ring normally never exceeds 30%, and usually is held to 17% maximum for dynamic applications or 25% maximum for static applications. Most O-rings should not be stretched more than 5% at installation. The radial clearance between moving components should be held to the smallest amount possible to prevent extrusion of the O-ring under pressure, yet not so small as to cause scoring of adjacent running surfaces against which the O-ring must seal.

5. Assembly Precautions

Sharp edges, such as threads, bore corners, splines, keyways or rough areas, that might cause the O-ring to be pinched or cut during assembly should be rounded, chamfered, masked, or protected by a sleeve. When installing the O-ring in its groove, avoid twisting it.

H. MATERIALS OF CONSTRUCTION

1. Sealing Face Materials

Most all mechanical face seals have a seal washer nose made of either impregnated synthetic carbon and graphite materials, or specially filled and processed thermosetting plastic compounds.

a. *Synthetic Carbons and Graphites*—Carbon and graphite materials can be very roughly divided into three main groups: hardened amorphous carbons, carbon-graphites and electrographites. Electrographites are of low strength, and have high coefficients of thermal conductivity. Hardened amorphous carbons are characterized by their higher strengths, and lower coefficients of thermal conductivity. Carbon-graphites have intermediate properties.

Although some aspects of the manufacture of carbon and graphite are covered by patents, the processes are still very much of an art, with each manufacturer keeping his own preferred procedures a secret. The mechanical and physical properties of carbons and graphites depend upon the way in which they are processed. These processing steps include choice, preparation, and the mixing of raw materials, the grinding of the cooled mixture to particle size for molding, the molding of parts, the baking of parts at elevated temperatures, and the impregnation of parts before or after graphitization at highly elevated temperatures for a period of days.

Carbons and graphites have good mechanical and thermal properties, especially at elevated temperatures, resist most chemicals except the most oxidizing acids, have good frictional properties, and are very stable dimensionally.

Friction and Wear Tester—Provides fast and precise evaluations of mechanical face seal material combinations, operating either dry or in liquids, under unit loads of 5 to 1,500 psi, velocities to 800 fpm and fluid temperatures to 400°F.

Carbon and graphite can be made with a wide variety of pore sizes, shapes and extents of interconnecting porosity. Most carbons from the baking process have a porosity of roughly 20%. The impermeability of these materials may be improved by cured liquid impregnations, metal impregnations, or formation of pyrolytic carbon in the pores. Liquid impregnation is normally carried out in an evacuated and sometimes heated tank containing the parts. A liquid impregnant, such as pitch, furfuryl alcohol, styrene, epoxy, or sugar solution is then flowed over and around the parts, covering them completely. A high pressure is applied to the tank and the liquid is forced into the evacuated pores. Molten metals, such as copper, lead, bronze, babbit, silver, cadmium, antimony, and nickel chromium, can be impregnated into the pores of carbon and graphite in much the same manner. Pyrolytic carbon can be deposited in the pores from a hydrocarbon gas mixture, such as methane mixed with nitrogen, being processed at atmospheric or lower pressures and highly elevated temperatures. Other impregnants used include barium fluoride, lithium carbonate, molybdenum sulfide, and TFE dispersions; these provide lubrication when operating in air, dry gases, and vacuum; also low temperature glasses to provide oxidation protection and lubrication at elevated temperatures.

The materials used to impregnate the pores of carbons and graphites generally determine the maximum operating temperature of the material; they also affect the material's hardness, strength, modulus, and coefficients of expansion, thermal conductivity, and friction. Many carbons and graphites resist temperatures to 600°F, and some can be used in air to 1200°F.

Chemicals which often cannot be adequately sealed by the use of carbon or graphite include bromine, hot liquid chlorine, chromic acid, chrome plating solutions, fluorine, hydrofluoric acid, iodine, nitric acid, and sulfuric acid in concentration of over 75%; the chemicals are usually more effectively dealt with by using a filled fluorocarbon material.

b. *Plastics*—Specially filled and processed thermoset resin-based seal face materials have proven to be economical materials for seals operating at lower pressures, speeds and temperatures, where sufficient lubrication is available. These materials can be molded to relatively close tolerances and intricate shapes, are free of interconnecting porosity, and have good resistance to many chemicals. They do, however, have relatively high coefficients of thermal expansion and low conductivity coefficients, plus PV rating limitations of around 10,000 for dry operation and 50,000 for lubricated service. Thermoset resin-based seal face materials are usually not recommended in applications with ambient temperatures above 400°F. Within these limitations, proprietary plastic seal face materials are often specified when sealing aromatic and chlorinated solvents, fuels, lubricating oils, mild alkalies and acids, fatty acids, ester plasticizers, antifreeze fluids and rust inhibitors, and detergent and water solutions.

2. Sealing Counterface Materials

The seal counterface, or opposing seal face, can be made of many materials, such as ceramic, iron, Ni-Resist, Stellite, tungsten carbide, bronze, stainless and tool steels, and various materials plated with dense chrome. In some applications, like or similar materials are run against one another, but in most applications better all-around performance can be obtained using dissimilar materials. The choice of which counterface material to use depends upon operating conditions, the configuration required, costs, and other requirements and limitations.

TABLE III

Coefficients of Thermal Expansion

Material	Coefficient (in/in/deg F)
Carbon and Graphite	$.8 - 4.0 \times 10^{-6}$
Alumina Ceramics	$3.1 - 3.7 \times 10^{-6}$
Low Expansion Nickel Alloys	$1.5 - 5.5 \times 10^{-6}$
Steatite	$3.3 - 4.0 \times 10^{-6}$
Ferritic Stainless Steels	$5.8 - 6.0 \times 10^{-6}$
Martensitic Stainless Steels	$5.5 - 6.5 \times 10^{-6}$
Gray Irons	6×10^{-6}
Malleable Irons	$5.9 - 7.5 \times 10^{-6}$
Alloy Steels	$6.3 - 8.6 \times 10^{-6}$
Nodular or Ductile Irons	$6.6 - 10.4 \times 10^{-6}$
Beryllium Copper	9.3×10^{-6}
Austenitic Stainless Steels	$9.0 - 10.2 \times 10^{-6}$
Aluminum Alloys	$11.7 - 13.7 \times 10^{-6}$
TFE Fluorocarbons	55×10^{-6}

3. Elastomers for Secondary Seals

One advantage of mechanical face seals over radial oil seals is that the elastomeric components of the former type perform secondary sealing functions that are relatively static or passive in nature; a radial oil seal, on the other hand, tolerates less degradation of its elastomeric member, which is the primary sealing mechanism. However, since the life of a mechanical face seal can be prematurely ended by excessive degradation of its elastomeric components, the compound used must be chosen with great care.

Assorted Rotating Seal Seat Designs

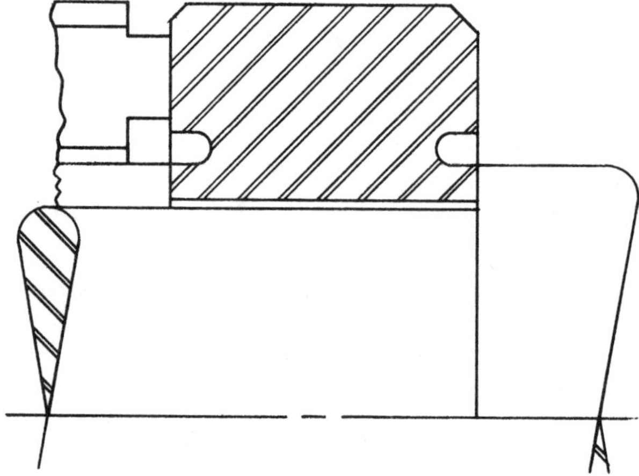

Fig. 2-35

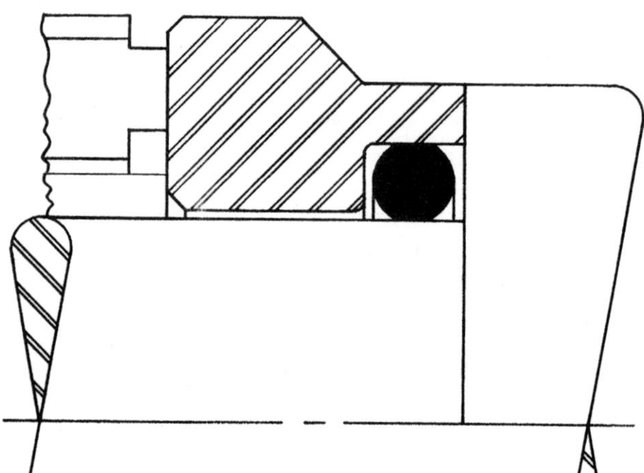

Fig. 2-36

Assorted Rotating Seal Seat Designs (Cont'd)

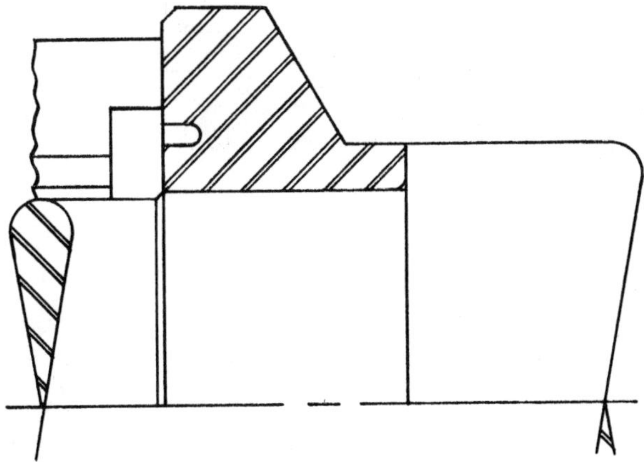

Fig. 2-37

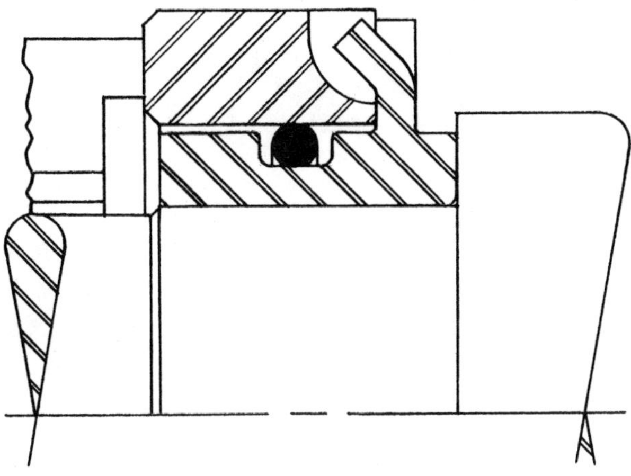

Fig. 2-38

Assorted Rotating Seal Seat Designs (Cont'd)

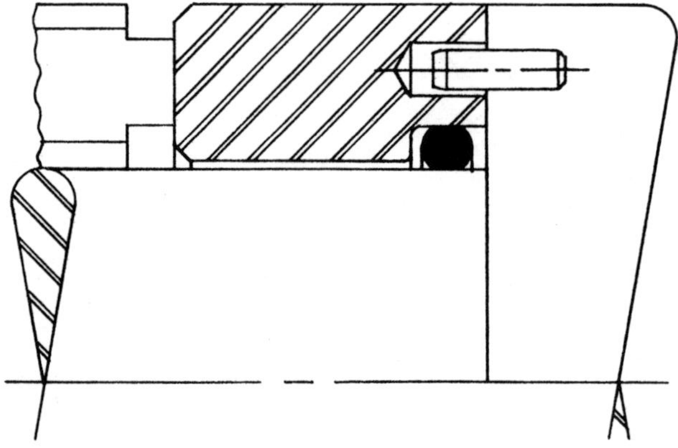

Fig. 2-39

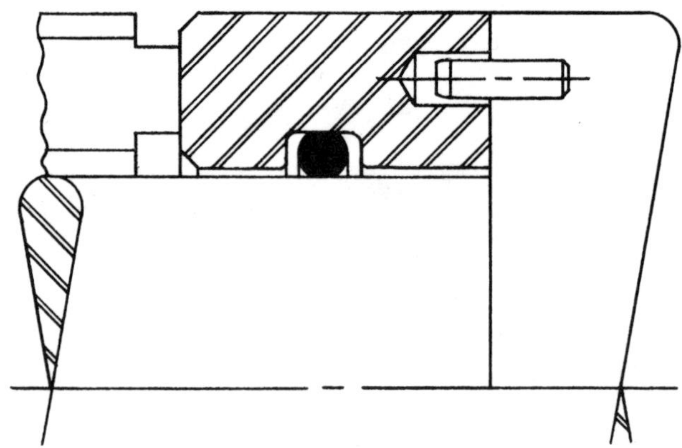

Fig. 2-40

Assorted Rotating Seal Seat Designs (Cont'd)

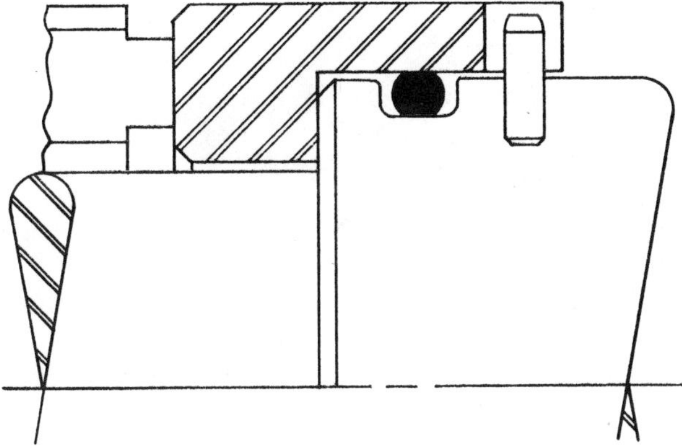

Fig. 2-41

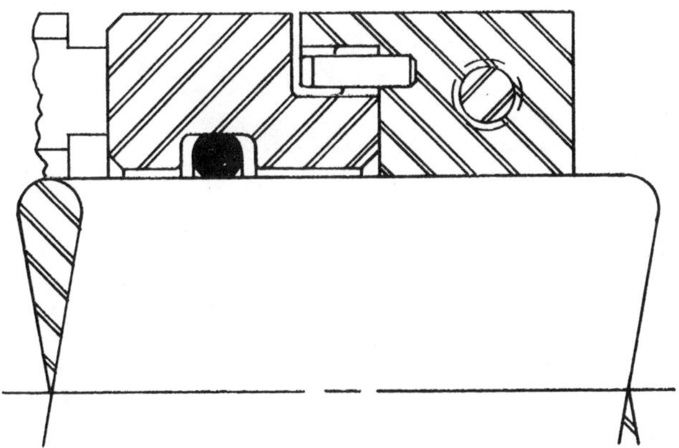

Fig. 2-42

Assorted Rotating Seal Seat Designs (Cont'd)

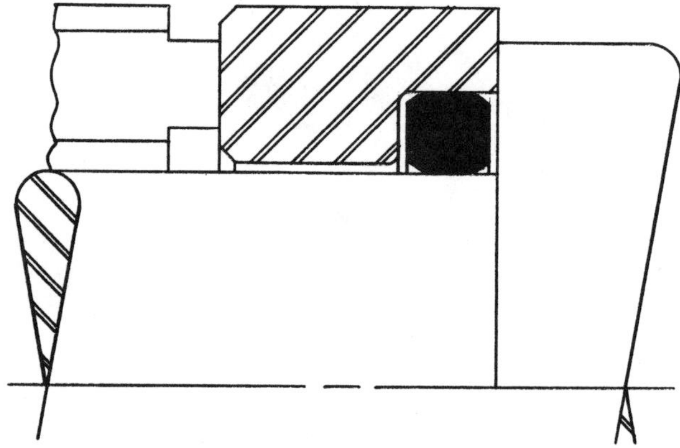

Fig. 2-43

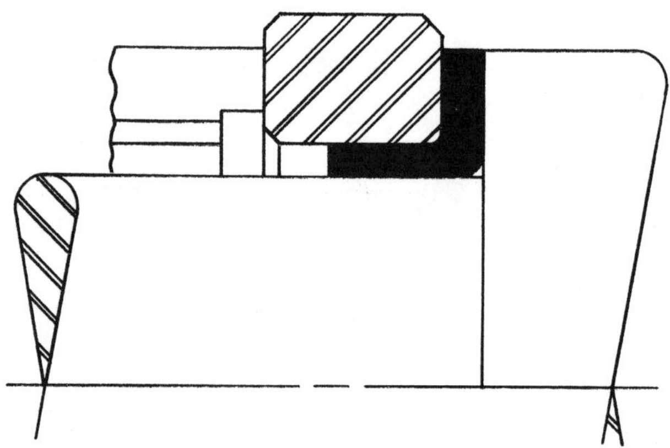

Fig. 2-44

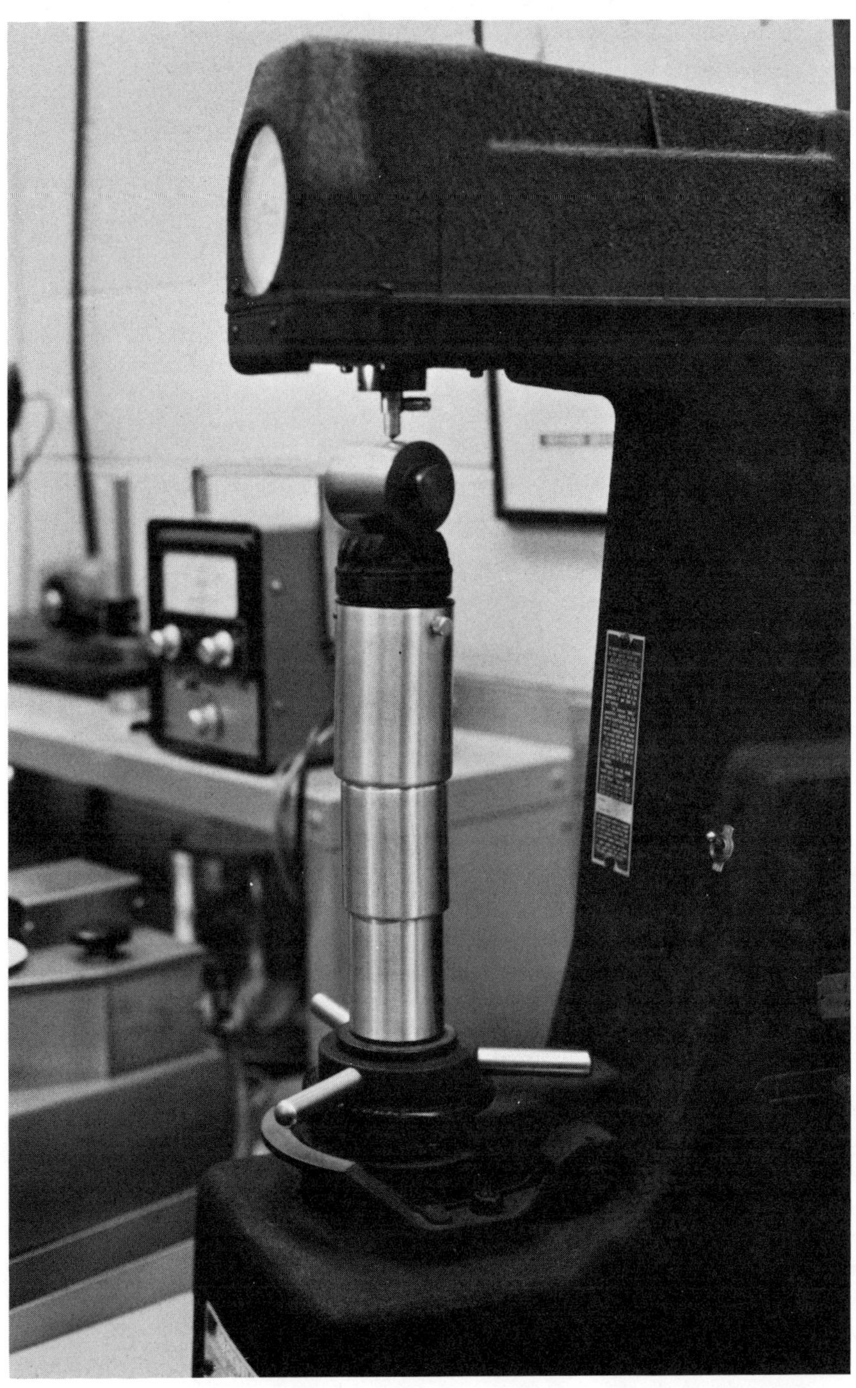

Rockwell Hardness Tester—This instrument is used to determine the hardness of various materials, based on the depth of penetration of a specified penetrator into the specimen under fixed conditions of test.

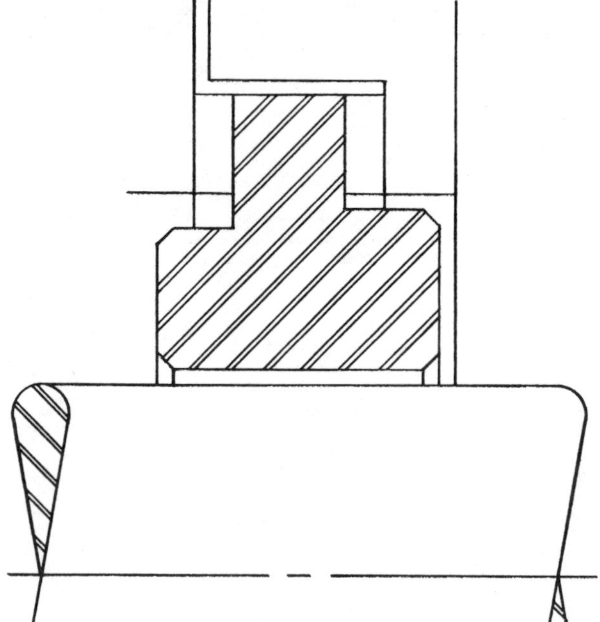

Fig. 2-45

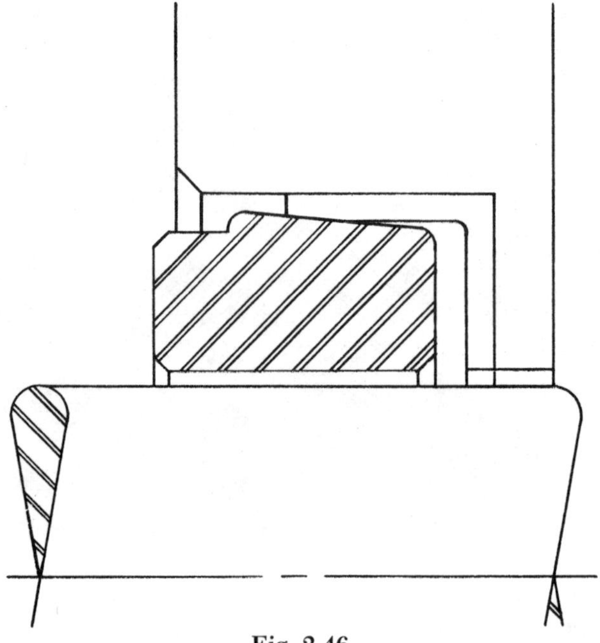

Fig. 2-46

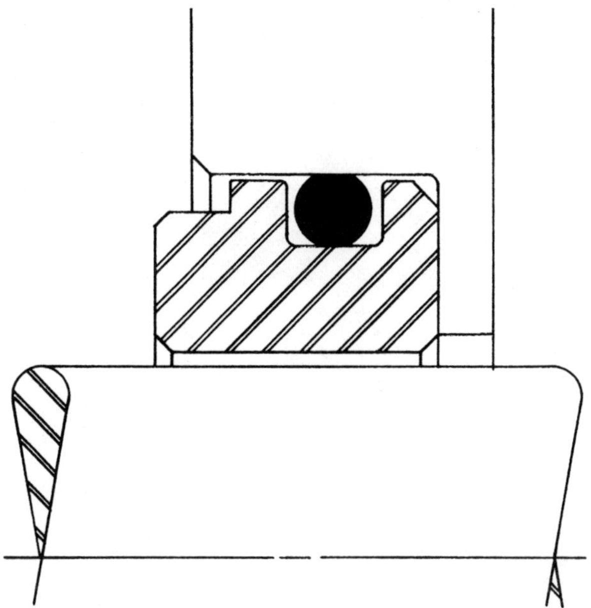

Fig. 2-47

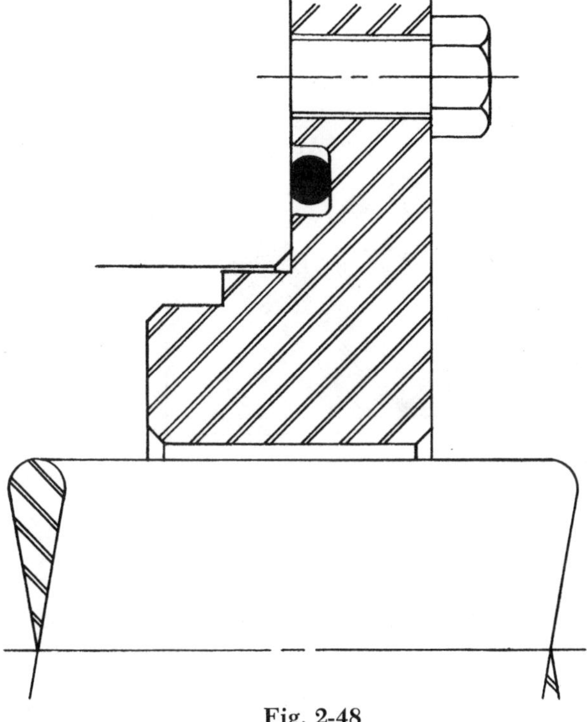

Fig. 2-48

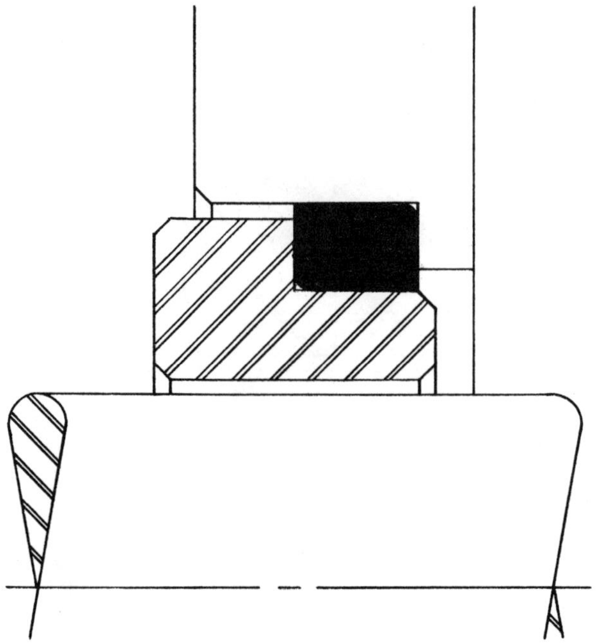

Fig. 2-49

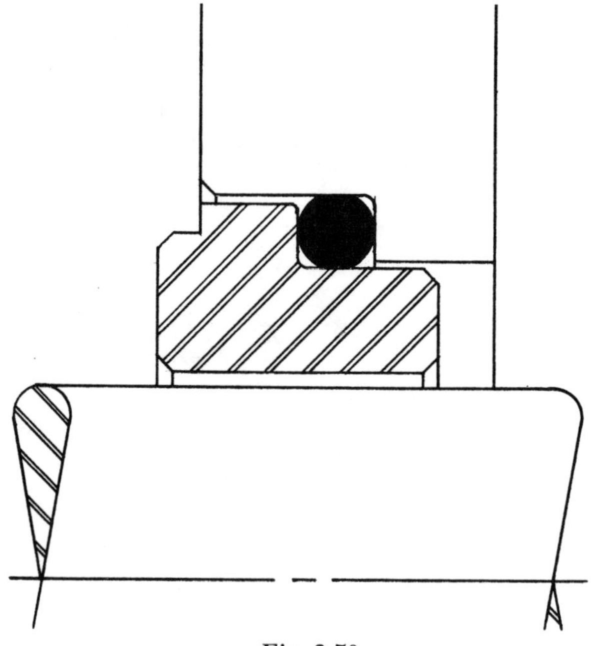

Fig. 2-50

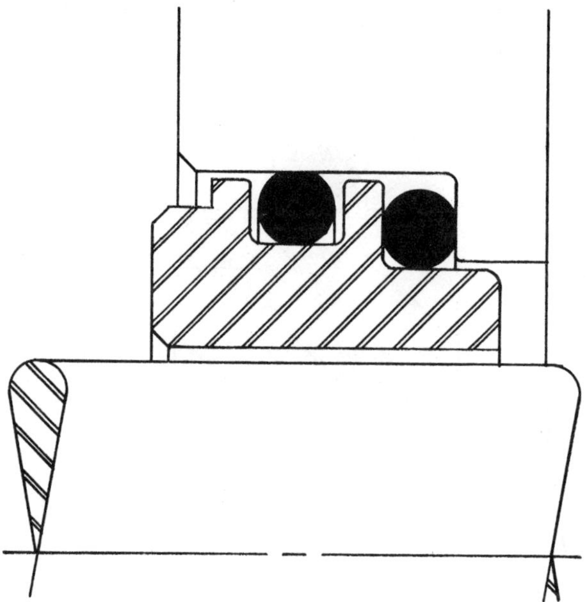

Fig. 2-51

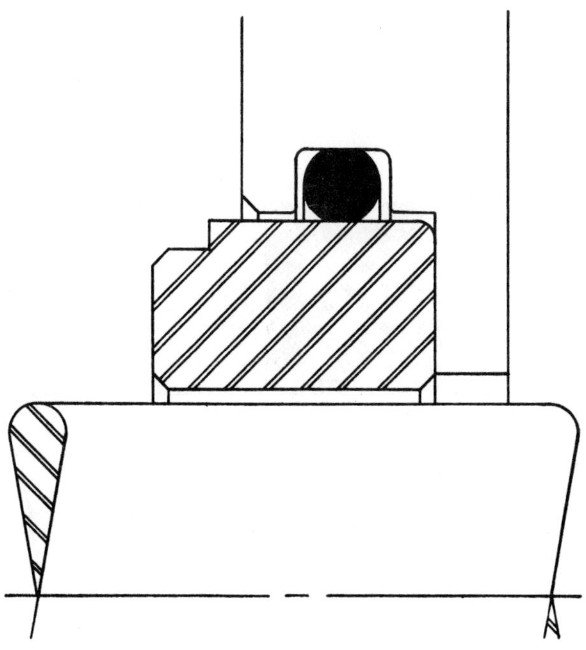

Fig. 2-52

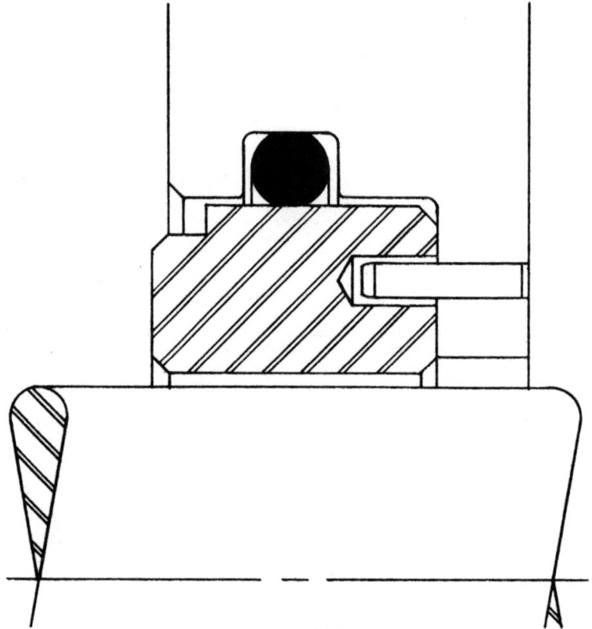

Fig. 2-53

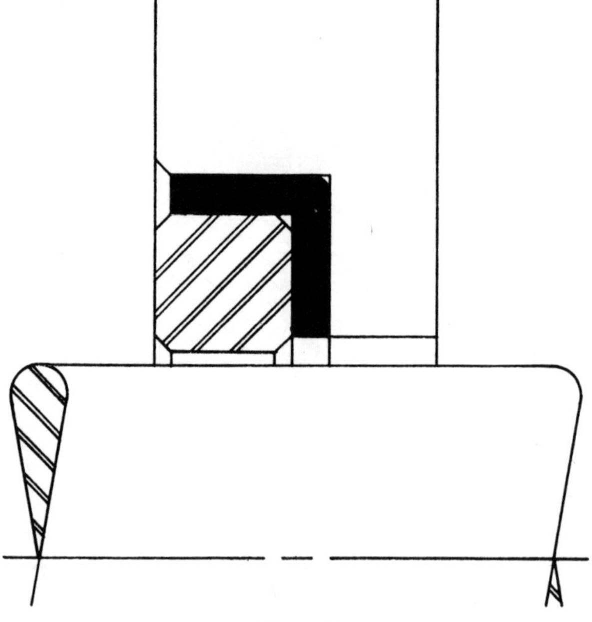

Fig. 2-54

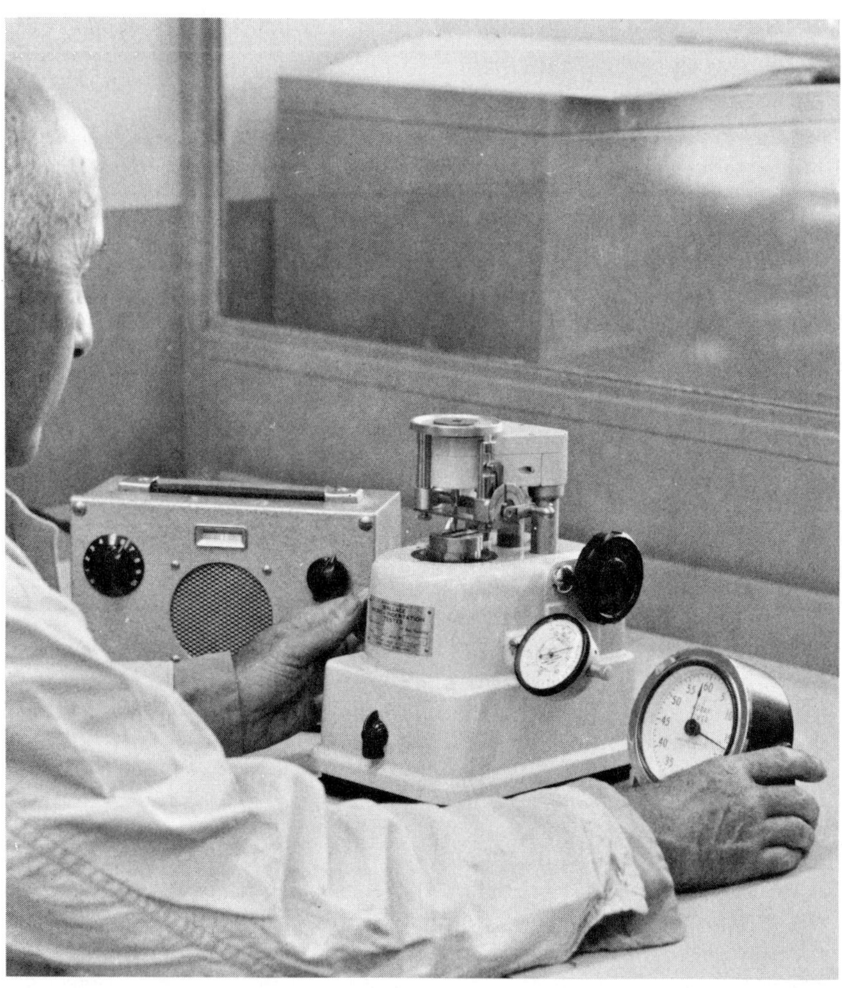

Wallace Micro Hardness Tester—Permits direct hardness measurements on small or thin section elastomeric parts. Equipped with visual and audio null indicators.

a. *Ethylene Propylene*—EP compounds can be used in environments ranging from −65 to 350°F. They have excellent resistance to heat degradation, low temperatures, sunlight, ozone, and weathering. Recommended for water, steam, phosphate ester base hydraulic fluids, and silicone oils and greases. Not recommended for petroleum base hydraulic fluids, oils, greases, gasoline, or di-ester base fluids.

b. *Chlorosulfonated Polyethylene*—These elastomers have excellent ozone resistance, good abrasion resistance, and excellent tensile properties. Operating temperature range of −40 to 250°F. Recommended for vegetable and some mineral oils, corrosive chemical environments, and water. Not recommended for aromatics and chlorinated hydrocarbons.

c. *Fluoroelastomers*—Very expensive elastomers, having exceptional chemical, compression set, and temperature resistance to over 450°F. Recommended for petroleum oils, halogenated hydrocarbons, silicone oils and greases, silicate ester base fluids, di-ester base fluids, and certain phosphate ester fluids. Not recommended for other phosphate ester base hydraulic fluids, low molecular weight esters and ethers, and applications requiring good low temperature flexibility.

d. *Silicone Elastomers*—Have excellent resistance to compression set, fatigue, ozone, sunlight, and weathering, and can be used in environments of −130 to over 450°F. Silicones have poor tensile strength and poor tear and abrasion resistance compared with most elastomers. Recommended for water, acids and bases, some phosphate ester base hydraulic fluids, EP-type lubricants and high aniline point petroleum oils, plus dry heat. Not recommended for low aniline point petroleum oils, gasoline, aromatic solvents, and silicates.

e. *Fluorosilicones*—These elastomers combine many of the properties of silicone with those of fluorocarbons, although their compression set resistance at elevated temperatures, and overall heat aging properties, are not as good as those of some other silicone compounds. Recommended for fuel, oil, and solvents, over a temperature range of −130 to 350°F.

f. *Nitrile*—Nitrile is a copolymer of butadiene and acrylonitrile. As acrylonitrile content increases, resistance to petroleum base oils and hydrocarbon fuels increases, as does strength, but low temperature flexibility decreases. To increase low temperature flexibility, it is usually necessary to sacrifice high temperature oil and fuel resistance, and strength. Recommended for petroleum oils and fluids, di-ester base lubricants, ethylene glycol base fluids, silicone greases and oils,

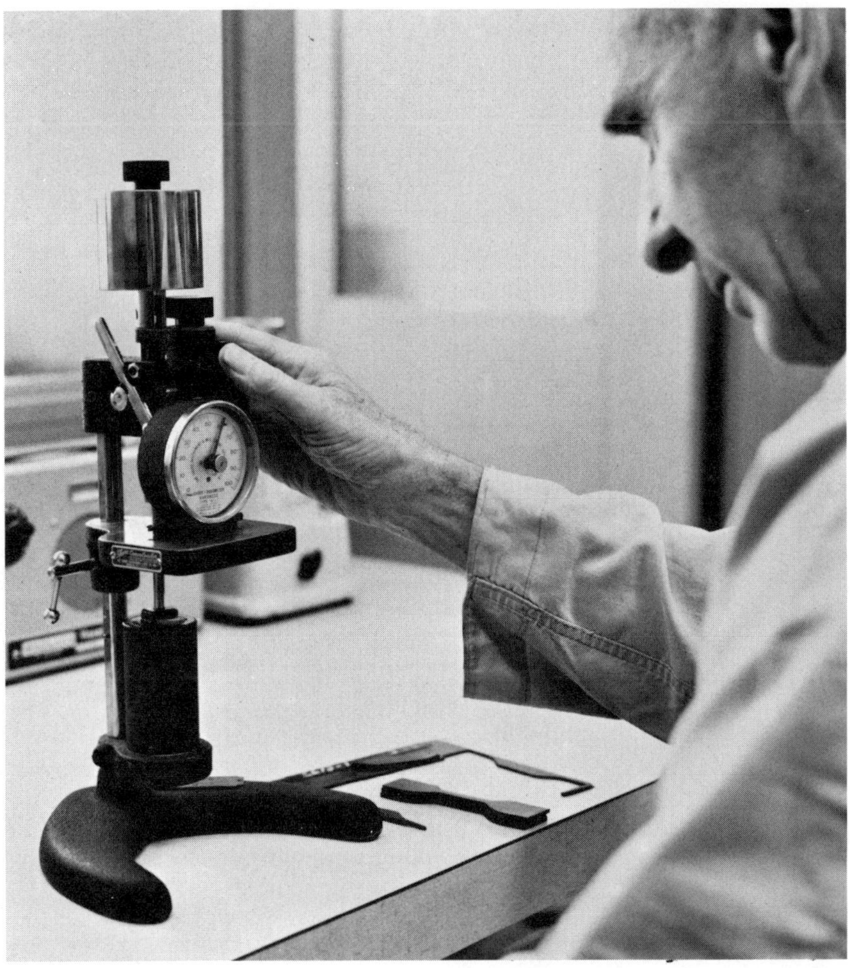

Durometer Hardness Measuring Instrument—A constant load device with an indentor, which penetrates into a surface, providing direct readings on a visual scale. Generally used for testing elastomeric plied specimens or samples 0.2 inch or thicker.

and water at temperatures ranging from −65 to 250°F. Not recommended for halogenated hydrocarbons, phosphate ester hydraulic fluids, or ozone service conditions.

g. *Carboxylated Nitrile*—This elastomer has a structural resistance to ozone, even after exposure to lubricating oils.

h. *Chloroprene*—This elastomer has good resilience, tear resistance, flex life, and good resistance to compression set, sunlight, ozone, and

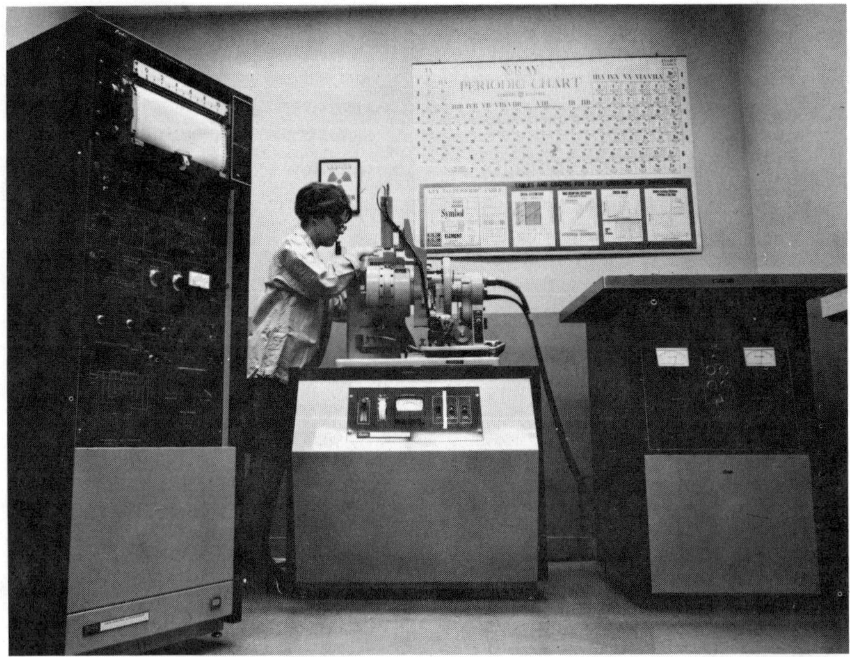

X-ray Spectrograph—Equipped with digital display and recorder, audio detector and direct graphic recorder. Samples of metals, plastics or elastomers are bombarded with x-rays, and reflections from an analyzing crystal are detected. The specific angles where responses are found provide element identification. The element range scanned includes aluminum (atomic number 13) to uranium (atomic number 92).

weathering. Recommended for high aniline point petroleum oils, nonaromatic gasolines, silicate ester type hydraulic fluids, and many Freon * gases at temperatures from −65 to 250°F. Not recommended for phosphate ester fluids, high aromatic content fuels, low aniline point oils, and oxygenated solvents.

i. *Isobutylene Isoprene*—Butyl type elastomers have good resistance to ozone, weathering, sunlight, and low temperatures. Until the introduction of EP type elastomers, butyl was the only elastomer which was satisfactory for phosphate ester type hydraulic fluids (fire resistant). In addition, its very low permeability to gases makes it particularly useful for vacuum applications. Operates at temperatures from −65 to 212°F. Recommended for phosphate ester type hydraulic fluids and silicone fluids and greases. Not recommended for petroleum fluids and di-ester based fluids.

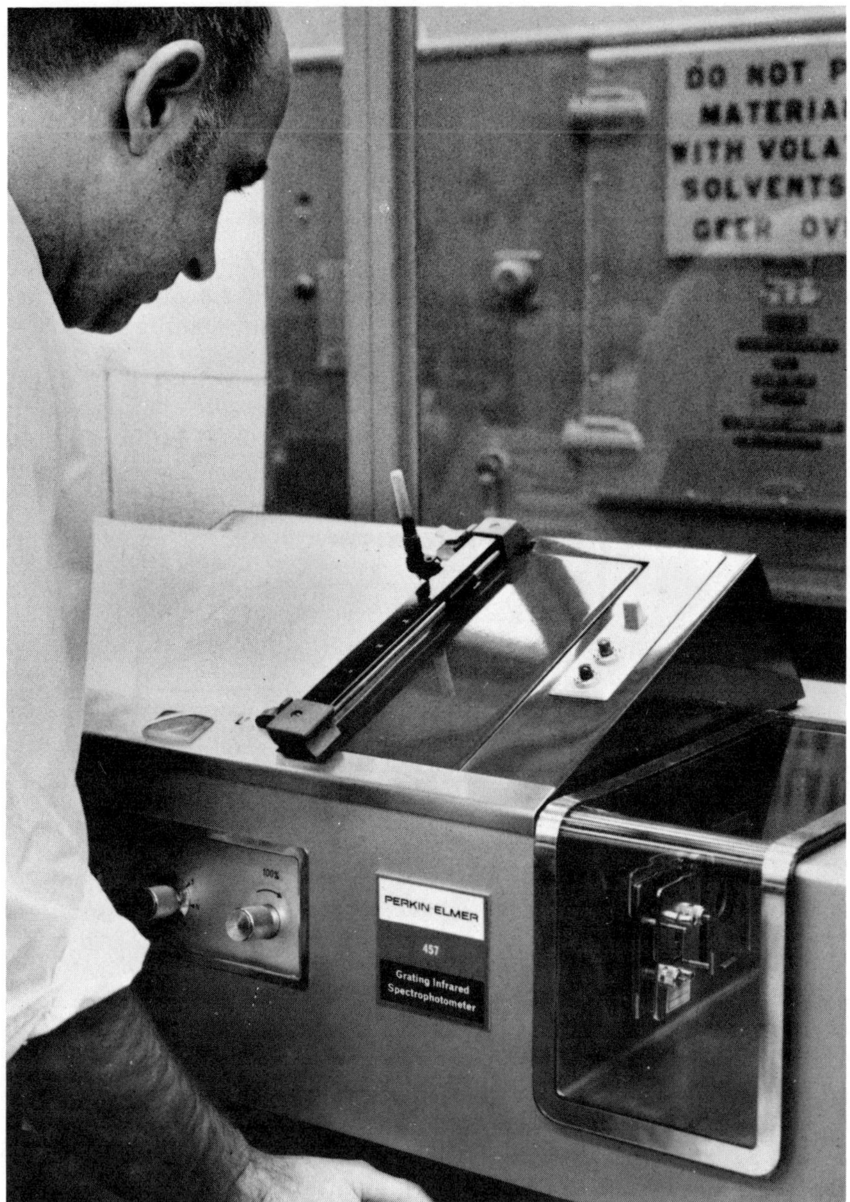

Infrared Spectrophotometer—This unit provides molecular structure data of various materials. Samples are prepared by various techniques and applied to crystals which are placed in the path of the infrared beam. The molecular excitation is graphically recorded, and results are compared to known standards. This machine yields data on many materials, principally organic in nature.

j. *Polyacrylic*—Polyacrylics are now used in many applications where nitrile was specified. These elastomers have excellent petroleum fluid, ozone and oxidation resistance and are serviceable in temperatures from 0 to 350°F. Recommended for most all petroleum base fluids, greases, and silicone fluids. Not recommended for water, steam, ethylene glycol, gasoline, alcohol, and strong acid or alkaline solutions.

k. *Epichlorohydrin*—As a class these elastomers show excellent resistance to ozone and have good heat resistance. Recommended for gasoline and petroleum-base hydraulic and lubricating oils. Not recommended for nonmineral base brake fluid.

l. *Polyurethane*—These polymers have high tensile strengths, but poor physical characteristics at temperatures over 180°F. Recommended for many petroleum fluids, aliphatics, and applications involving gamma radiation. Not recommended for acids, alkalies, hot water, steam, and certain oil and grease additives.

4. Fluorocarbon Resins, TFE and FEP

Filled and nonfilled fluorocarbon resins are increasingly used in mechanical face seals as the material for construction of seal faces, secondary seals, gaskets, and restriction bushings. TFE-fluorocarbon resins are tetrafluoroethylene polymers, usually fabricated by cold-forming and sintering techniques. FEP-fluorocarbon resins are tetrafluoroethylene-hexafluoropropylene copolymers, usually fabricated by melt extrusion or injection molding. Most of the important end-use properties of TFE and FEP resins are equivalent, the principal difference being that FEP resins have a somewhat lower continuous service temperature (400°F, compared with 500°F for TFE resins). Up to their respective upper use temperatures only very few chemicals react with these resins, for example, molten alkali materials such as sodium, turbulent liquid or gaseous fluorine, and a few fluorochemicals including chlorine trifluoride (ClF_3) or oxygen difluoride (OF_2) which readily liberate free fluorine at elevated temperatures. To a minor degree, halogenated organic chemicals may be absorbed by fluorocarbon resins; FEP resins are less affected in this way than TFE resins.

Fluorocarbon resins have exceptionally low friction in nonlubricated applications, especially at low surface velocities and pressures higher than 5 psi. The coefficients of friction increase rapidly with sliding speeds up to about 100 feet per minute, under all pressure conditions. This pattern of behavior prevents "stick-slip" and its resultant squealing. Above 150 feet per minute, sliding velocity has relatively

Gas Chromatograph—Analytical device for quantitative and qualitative analysis of polymers and compounds, with graphic recorder. Indicates separate constituents by chart comparison to known standards.

little effect at combinations of pressure and velocity below the material's PV limit. The PV and wear values of fluorocarbon resins and compositions are strongly influenced by ambient operating temperatures. Fluorocarbon resins, like many plastics, have a tendency to cold-flow under load or compression, but this can be offset by proper design, and improved by the additions of fillers. PV and wear values can also be improved by the addition of various fillers such as graphite, bronze, carbon, asbestos, zirconium, glass fibers, molybdenum disulfide, coke flour, copper, and ceramics.

5. Metal Components

Stamped, formed, machined, cast, and powdered metal components can be supplied in a diversity of materials including plated steel, stainless steel, Hastelloy, brass, aluminum, monel, titanium, and zinc.

Specific material recommendations for particular operating environments may be found in Section X. To effect cost savings, however, plated carbon steel can be used in many applications instead of the more expensive stainless steel materials listed. Following are some of the platings commonly used on carbon steel components for mechanical face seals.

a. *Cadmium*—Silver-gray in appearance, cadmium offers excellent corrosion resistance in salt atmosphere. This plating is smooth and uniformly deposited with good appearance; it adheres well to the base metal, provides a suitable surface for post-plate treatments, and is economical. Cadmium plating, however, is soft, not resistant to abrasion and has limited porosity resistance under .0002-inch plate thickness. Moreover, it is poisonous, and therefore it cannot be used in food processing and packing equipment; and its use is controlled by U.S. government priority during national emergencies. Cadmium-plated parts should not be used in applications which reach or exceed temperatures of 450°F.

b. *Cadmium and Chromate*—Supplementary chromate treatments reduce the oxidation of the cadmium base plate by sealing the pores with a protective film. Supplementary chromate treatments are damaged if used on cadmium-plated parts which are continuously exposed to temperatures in excess of 150°F, or intermittently exposed for short periods to temperatures exceeding approximately 300°F. Chromate treatment can be applied in red, blue, green, black, bronze, olive drab, yellow iridescent, or clear for color identification.

c. *Zinc*—An economical substitute for cadmium, zinc has a bluish-white appearance and requires a thicker plating to equal the corrosion resistance of cadmium. Zinc has limited porosity resistance under .0002 thickness and is not recommended for use in a heat environment exceeding 500°F.

d. *Zinc and Chromate*—A supplementary chromate treatment offers the same advantages and limitations when applied over a zinc plating as it does for a cadmium plating as described in paragraph (b) above.

e. *Black Oxide*—Provides a uniform dull black, or luster black appearance when coated with oils. Provides mild corrosion and rust resistance economically. Black oxide is a uniform and adherent film which withstands severe deformation and does not effect the dimensions of a part, but it is not generally recommended for exposure to moisture unless coated with oil.

f. *Iron Phosphate*—Provides a uniform gray finish having mild rust- and corrosion-resistance when treated with oils. A uniform and adherent film which is temperature resistant to 250°F.

Rockwell Superficial Hardness Tester—This machine is designed for testing thin sheet metal, nitrided and lightly case-hardened steel, and for any hardness tests requiring exceptionally shallow indentations.

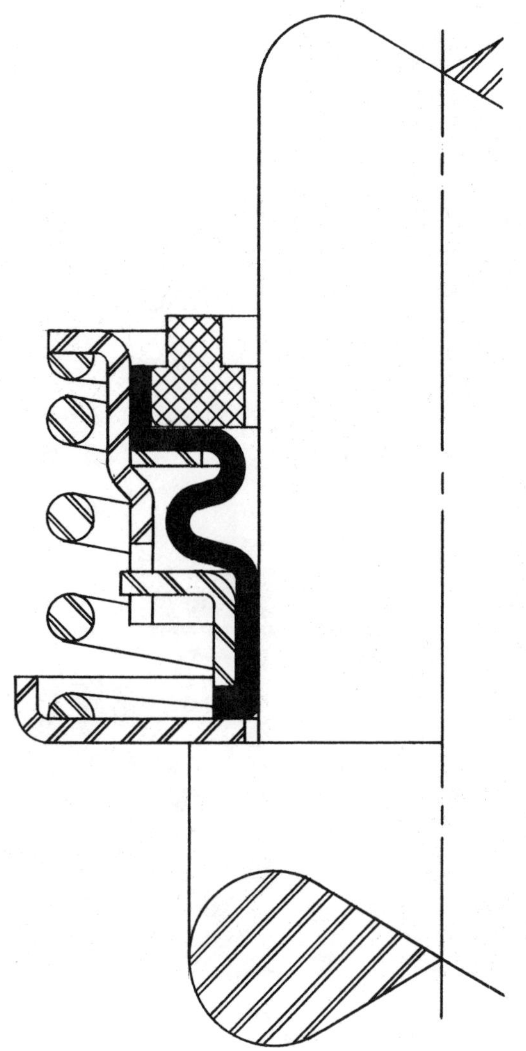

Fig. 2-55 Molded

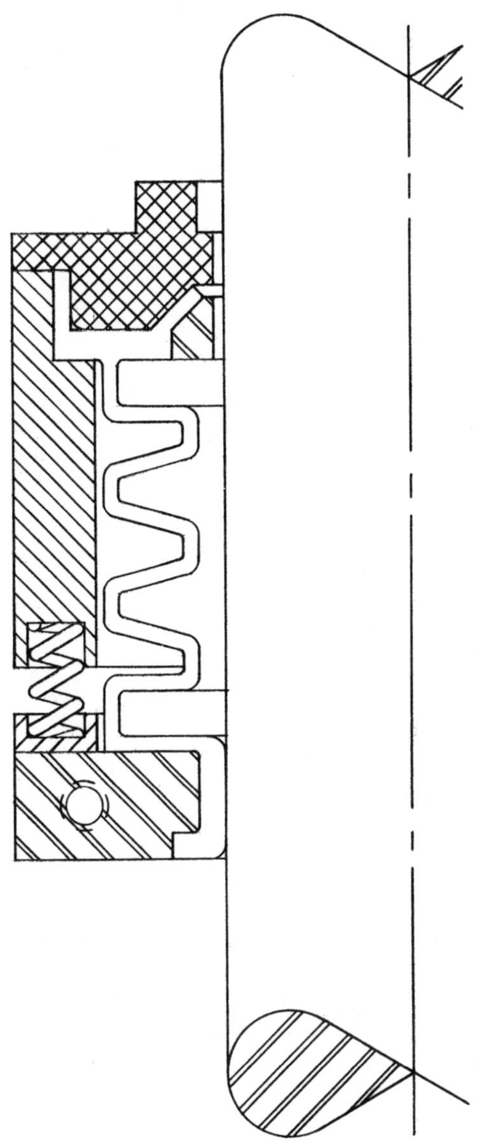

Fig. 2-56 Machined

Bellows Secondary Seals (Cont'd)

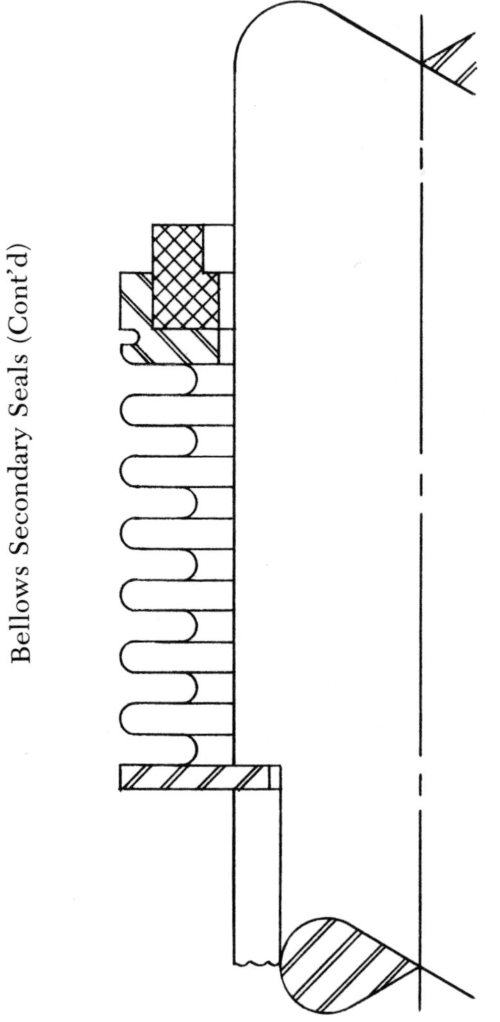

Fig. 2-57 Formed

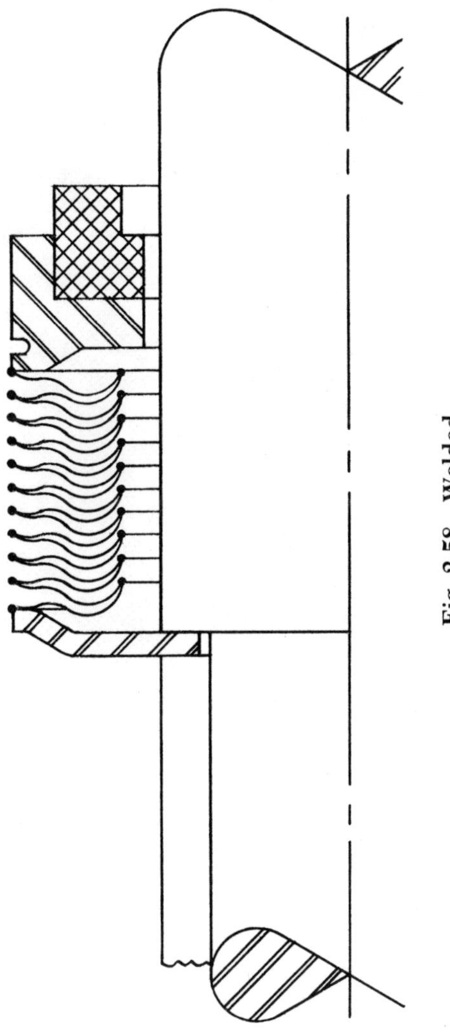

Fig. 2-58 Welded

Optical Comparator and Measuring Machine – Collimated illumination projects outlines with 10× to 100× magnification onto 30-inch display screen for obtaining accurate measurements of small, complex, or difficult to measure parts, without using physical means. Electronic coordinate table provides direct line positioning and measuring control, in any direction, to .0001 inch. Electronic instant base zero for actual dimensional readings at any location, through digital display and direct printout device.

Slideable Secondary Seals

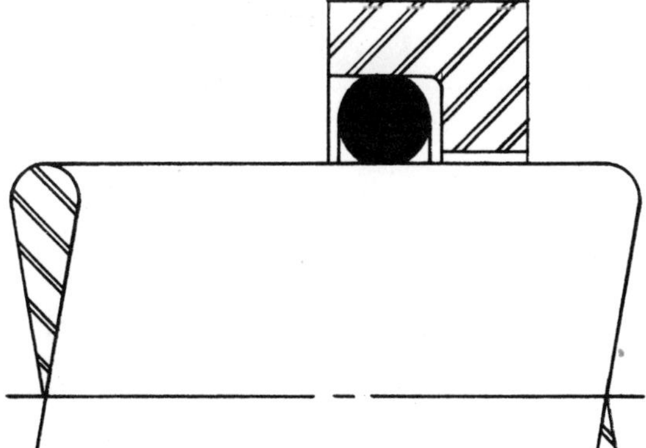

Fig. 2-59 O-Ring

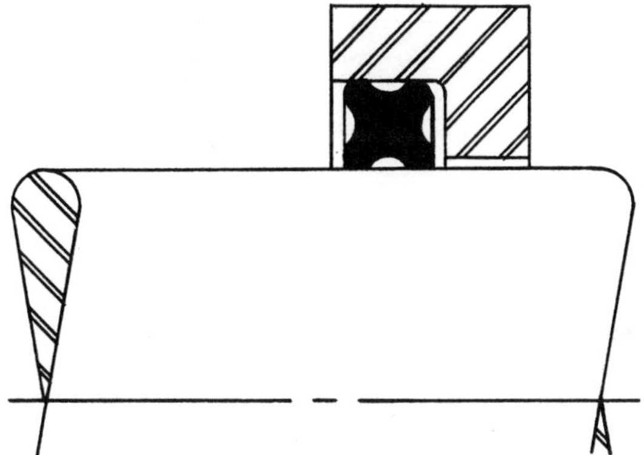

Fig. 2-60 Lobed Ring

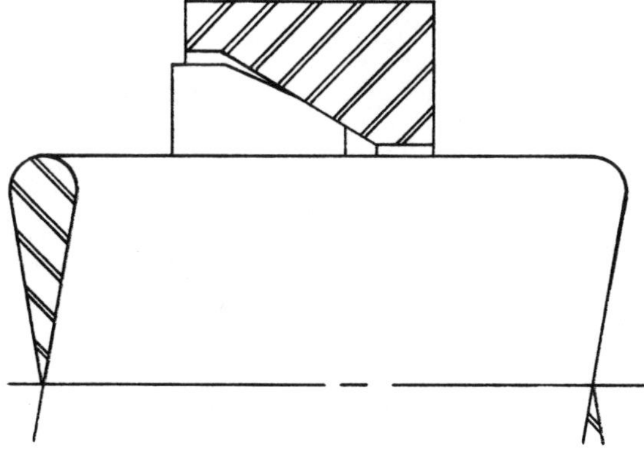

Fig. 2-61 Wedge

Slideable Secondary Seals (Cont'd)

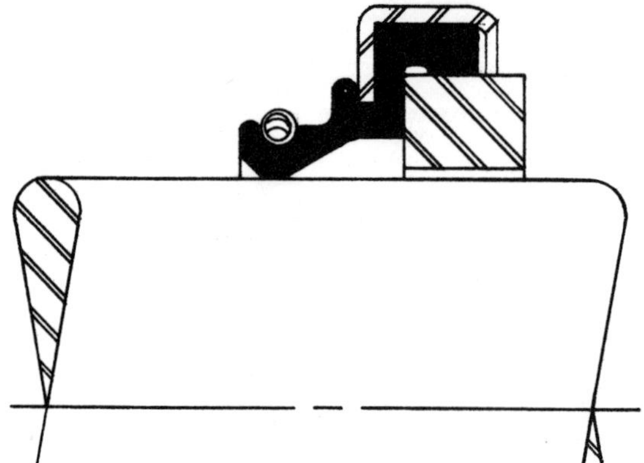

Fig. 2-62 Lip Seal

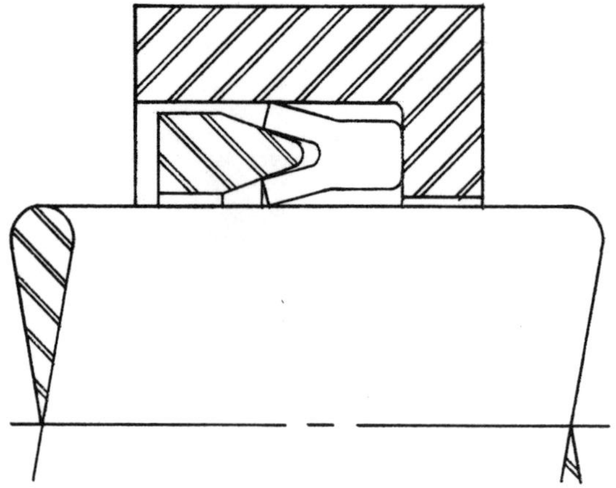

Fig. 2-63 "U" Cup

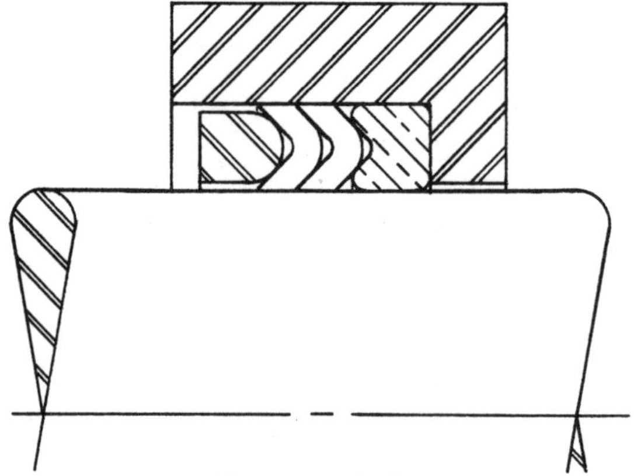

Fig. 2-64 "V" Ring

Radial Wall Measuring Fixture and Optical Comparator—For checking uniformity of wall thickness; that is, concentricity of inner and outer cylindrical surfaces. Accurate to .0005 inch, using a 20× lens in the optical system.

3
ENVIRONMENTAL CONSIDERATIONS

Following are some of the major causes of inadequate seal performance. It is very important that seal manufacturers are informed of such conditions if they exist or are expected.

A. ABRASIVES

Even though mechanical face seals are usually effective in dealing with abrasive conditions, abrasives should be kept to a minimum. When a mechanical face seal is sealing in a dynamic mode, the mating faces which effect the primary pressure seal are normally separated by a lubricating film of the fluid being sealed, only 1 to 5 microinches thick. Abrasive particles in the range of 1 to 10 microinches in this fluid will cause increased wear at the sealing surfaces which may eventually result in excessive leakage or premature failure of the seal. There are several sources and types of abrasives:

1. Foreign matter in the sealed system, such as dirt, casting sand, etc.
2. Products of corrosion in the system, such as rust scale.
3. Inherent characteristic of the fluid to be sealed, such as in a slurry.
4. Solid formed in the sealed system or between the sealing faces due to degradation, decomposition, or crystallization.
5. Wear debris from mechanisms, such as gears, cams, and bearings, inside the sealed system.
6. Abrasives outside the sealed system, such as sand and dirt, on the atmosphere side of the seal.

The amount of abrasion resulting from particles in the sealed fluid depends upon the hardness, size, shape, toughness, and number of particles. Additives can reduce or prevent the formation of abrasive particles in lubricating fluids. There are also polar antiwear additives that can reduce the abrasive effects of such particles by dispersing them and by preventing them from adhering to one sealing face, where they act as cutting tools against the opposing face.

Seals which must cope with excessively abrasive conditions are often provided with tungsten-carbide faces, which are generally brazed on or press-fitted into metal structural members of the seal. The tungsten-carbide is generally of two differing grades, one being slightly harder than the other. This method of dealing with abrasives is ordinarily limited to seals with face speeds less than 2,000 fpm and sealing pressures less than 50 psig, and for sealing liquids where the solids content is less than 20% by volume and where the abrasives are of a relatively soft nature. Other methods of dealing with abrasives are covered in Section IV.

B. HEAT

Excessive heat can cause a rapid and sometimes violent vaporization of the fluid film between the sealing faces, resulting in severe wear, noise, seal leakage, or even damage to or destruction of critical seal components. Such heat may also cause coking of oils being sealed, thus building up abrasive deposits around the sealing faces. With some fluids, temperatures must be controlled in the seal area to minimize solidification of the fluid at the mating faces. Very high temperature may result in heat checking of some seal face materials; and quick changes in temperature may induce thermal shock or cracking in others. High temperatures also tend to degrade elastomeric materials used in secondary sealing members and to increase the corrosion rates of metal seal components. Methods of dealing with heat are covered in Section IV.

C. DRY OPERATION

Regardless of materials of construction and seal design, any mechanical face seal, even those designed for "dry face" operation, will exhibit some or all of the following tendencies when operated quasi-dry or dry, rather than with a fluid lubrication film present at the sealing interface.

1. Increased start-up and running friction and torque, and increased heat output.
2. Accelerated wear and damage of the dynamic sealing surfaces, such as blistering, cracking, chipping, grooving, pitting.

3. Audible squealing or ringing at times.
4. Increased wear of anti-rotation mechanisms of the seal.
5. Increased degradation of secondary seals.
6. Increased build-up of deposits on and around the seal.
7. Increased leakage, due to thermal and mechanical distortions, wear and damage of sealing surfaces and components, restricted axial compliance of components, build-up of deposits between and around the sealing faces, and seal fluttering and outgassing due to interfacial film vaporization.
8. Shortened seal life.

Unless a mechanical face seal has been designed for such service, running without lubrication can cause rapid failure of the seal. Seals may run dry because of a vacuum in the sealed system, flashing of the fluid at the sealing face, air trapped in the sealed system, or cavitation. More information on dealing with dry operation may be found in Section IV.

TABLE IV

Lubricating Qualities of Mediums Sealed

Medium	Lubricating Quality
Liquid Nitrogen	Very poor
Hot Water and Steam Water and Detergents "Freon" 12,22,114 Carbon Dioxide Oxygen-Argon Helium-Hydrogen Nitrogen High Vacuum Liquid Oxygen Liquid Hydrogen Liquid Nitrogen Tetroxide	Poor
Liquid Halogens Gasoline Kerosene	Poor to fair
Acids Bases Neutral or Inert Atmosphere Reducing Atmospheres Air	Poor to good
Petroleum Lubricants Synthetic Lubricants Greases (except Lithium Soap Base)	Good

4
ENVIRONMENTAL CONTROL

The primary objective of environmental control is to maintain a clean liquid film between the dynamic sealing faces of a seal, and to maintain favorable ambient temperatures. Although such control is not practical in many situations, it should be appreciated that the life and performance of most seals can be greatly improved thereby. Following are some of the more common methods of controlling the environment in the vicinity of the seal.

A. FLUSHING

In sealing a pump, or a system containing a pump, flushing is used:

1. To cool the seal face to prevent the interfacial lubricating film from flashing or vaporizing;
2. To prevent highly volatile liquids, such as ammonia and gasoline, from forming a vapor trap around the sealing faces;
3. To prevent sediments and polymers from accumulating;
4. To provide a fluid barrier to keep abrasives from reaching the sealing faces.

Flushing of the seal cavity is accomplished by tapping the discharge side of a system and piping the fluid into the seal chamber, often through cooling or filtering units.

B. QUENCHING

Quenching may be accomplished by providing a bottom-vented quench gland external to the seal cavity and introducing water into the top of the gland, flooding the shaft immediately adjacent to the exterior of the seal. Leakage of water around the shaft to the outside of the quench gland, instead of out the bottom vent, is minimized with a throttle bushing, auxiliary packing, or radial lip seal.

The process of quenching is often used to carry away any products that leak past the seal or to prevent their contact with the atmosphere; it is therefore used as a safeguard when sealing gases and strong oxidizing and corrosive acids. Quenching also cools external seal components, provides seal interface lubrication under vacuum or suction lift conditions, and prevents certain fluids from crystallizing into abrasives on the outside of the seal.

C. JACKETING

In order to prevent liquids from vaporizing or solidifying it is at times necessary to heat or cool the seal cavity area. This can be accomplished by circulating fluids through a jacket around the seal cavity or around the stationary seal seat member.

D. DEAD-END FACE LUBRICATION

When sealing liquids with poor lubricating properties or containing mild abrasives, or in applications where the sealing faces would other-otherwise run dry, it is often advantageous to provide forced dead-end lubrication. This is accomplished with a special stationary sealing ring ported to allow lubricants to be introduced at high pressure but in very small amounts directly into the dynamic sealing interface. The lubricant used is usually a high grade oil or grease.

E. CIRCULATING FACE LUBRICATION

Preferred to the above method when cooling or heating is desirable or when sealing more severe abrasives, forced circulating lubrication is accomplished with a special stationary sealing ring ported at two places, these ports leading to a groove confined within the dynamic sealing interface. A lubricant, usually water or oil, is then circulated in through one port, around the groove in the dynamic sealing interface, and out the second port.

F. GREASE PACKING

Another method of dealing with mediums containing abrasives is grease packing; like dead-end face lubrication, this can be accomplished without piping fluids. In this method of environmental control, the mechanical face seal sealing the fluid medium is installed at one end of a gland, the other end being sealed by a radial shaft seal or felt, and the seal gland is filled with grease, which may or may not be replenished at regular intervals. The grease lubricates the mechanical face seal and prevents the abrasive-containing medium from entering the dynamic sealing interface. Some silicone greases are suitable for temperatures ranging from −100 to 450°F.

5
AUXILIARY EQUIPMENT

The following auxiliary equipment can be used to help provide a fluid of controlled cleanliness, temperature, pressure, or flow rate for use in flushing, quenching, jacketing, or force-lubricating mechanical face seals, in order to provide the seals with a favorable operating environment.

A. CYCLONE SEPARATORS

When the solids in a process stream are heavier than the liquid itself, do not constitute more than 10% by weight of the fluid being pumped, or 25% by volume, are not excessive in size, and if the fluid viscosity does not exceed 25 centistokes, cyclone separators offer an effective, reliable, and low cost means of removing the particles from the fluid.

Normally used with pumps, the separator is located on a hydraulic line running between the suction and delivery side of the pump, with a second line running from the clean liquid discharge port at the top of the separator to the seal being flushed, quenched or forced circulation lubricated.

The cyclone separator is a conical chamber with the small end at the bottom. At the side, and tangential, there is the inlet pipe for raw fluid; there are two outlet pipes, one for clean fluid near the top, and one for dirt discharge at the bottom.

The particle-laden liquid enters the cone and flows in a vortex because the inlet is tangential. The particles are driven by the rotation to the side of the cone and thence to the bottom outlet; the cleaner fraction of the liquid is permitted to flow out the top.

The dirt discharge, about 20% of the total input volume, contains 80% or more of the particles, down to as fine as 2.5 microinches.

B. PRESSURIZATION UNITS

Self-contained pressurization units are commercially available in a variety of sizes and capabilities to circulate sealing liquids, usually lubricating oils, through glands sealed at both ends by mechanical face seals, at controlled temperatures and pressures. When circumstances justify the cost, these units provide an effectively controlled environment for mechanical face seals operating under highly detrimental conditions, or for those installed in equipment which is critical, expensive, or difficult to service.

C. HEAT EXCHANGERS

These devices offer an effective means of maintaining desired temperature in the seal area; they are commercially available from a number of manufacturers.

D. AIR COOLERS

A low-cost device used in providing a cooler environment in the seal area. These low-cost units are simply finned convectors for dissipating excess heat; they require little or no maintenance other than occasional cleaning of the fins to remove accumulated dust so that radiating efficiency is maintained.

E. ROTAMETERS AND FLOW-RATE CONTROLLERS

A rotameter is a gauge consisting of a graduated glass tube containing a free float for measuring the rate of flow of a fluid. A flow-rate controller is a type of adjustable needle valve used to control the rate of flow of a fluid. When used in conjunction, such devices are useful for regulating the rate of fluid flow to seals being flushed, quenched, or lubricated by forced circulation, provided that the supply pressure is relatively constant. Where supply pressures fluctuate, the use of a pressure-compensated flow-control valve is recommended for a more constant delivery volume.

F. PRESSURE SENSITIVE SWITCHES

In order to avoid damage to mechanical face seals because of running dry, pressure sensitive switches are often installed in a system in order to shut down operation of a unit, or actuate audible or visual alarms if a pressure drop in the system occurs. Such switches can also be used to sense excessive pressures in a like manner.

G. STRAINERS AND FILTERS

Often installed in lines circulating flow to seals being flushed, quenched, or lubricated by forced circulation, strainers or filters are often recommended, especially during plant commissioning or start-up of units after a period of shut-down, in order to remove pipe scale and dirt from the fluid immediately before the fluid reaches the seal area. However, such strainers and filters must be removed after start-up unless a proper maintenance routine, which includes the thorough cleaning or replacement of these devices, can be assured, or unless pressure sensitive switches are installed to warn of excessively dirty strainers or filters.

H. LEAKAGE DETECTORS

A commonly used leakage detector consists of a container fitted with dipstick, float, and switch, and a bottom-mounted drain. The container must be installed below the seal or seals being monitored since the system relies on gravity feed of leakage from the seal or seals to the detector. When the accumulated leakage reaches a predetermined height in the container, the raised float engages the switch, which can actuate alarms, shut down operation, and perhaps start a stand-by unit.

6
USING OPTICAL FLATS AND HELIUM LIGHT SOURCES

Face flatness is usually inspected and measured by the use of monochromatic light and an optical flat. The monochromatic light, produced by an electric discharge through low-pressure helium, has a dominant wavelength of 23.2 microinches. The optical flat is made of Pyrex or of transparent fused quartz of optical quality, with the reference face lapped and polished flat to within 1 to 5 microinches.

The flat and the seal face to be checked must be clean, dry, and reflective; if the seal face has a dull or matte finish, it may be polished for inspection by gently rotating the part in a figure-eight pattern on a fine grade of polishing paper stretched taut and wrinkle-free over a polishing stand stage which has been lapped to a high degree of flatness. Both the part and the flat should be at room temperature.

The flat is placed upon the surface to be inspected, reference face down, and is illuminated by the helium light. The combination should be viewed perpendicularly from a distance of at least six times the diameter of the part. Light and dark bands will be seen across the face of the piece.

These bands are analogous to contour lines applicable to the face being tested relative to the optical flat; that is, to the extremely thin gap between the two surfaces. They are the result of optical interference between light rays reflected from the flat and from the seal face. If the bands are straight and evenly spaced, the seal face is flat; spacing between the bands becomes greater as the optical flat and the seal face become more nearly parallel. Each band corresponds to a half wavelength (11.6 microinches) difference from the adjacent band. Curvature and uneven spacing of the bands indicate lack of flatness. The amount the bands curve, with reference to this distance between them, indicates the amount of deviation from perfect flatness.

Optical Flat and Helium Light Source – For checking seal face flatness.

7
HANDLING AND INSTALLATION

A. GENERAL

Rough handling of mechanical face seals can scratch and chip the sealing faces and perhaps destroy their flatness as well. Unequal thickness of a gasket or of applied gasket cement, and excessive or uneven seal-gland bolt tightening can cause misalignment or distortion of critical parts. Sharp corners on bores, shafts, and sleeves, and burrs raised by set screws, can cut secondary seals. Sealing face surfaces may also be contaminated with dirt and other foreign matter if care is not exercised. All of the above conditions can cause premature failure of the seal.

If the seal is likely to run dry during the initial start-up period, the sealing face may be lubricated before installation with the fluid to be sealed, provided this fluid is a good nonabrasive lubricant. (See Table IV.) Should rather long initial dry running be expected, the sealing faces should be prelubricated with oil in most cases. Do not use wiping materials which will leave any lint or fibers on or around the sealing faces. Often, shafts and bores should be lightly lubricated prior to installing seal components to insure proper seating.

Before starting up mechanisms with newly installed seals it is recommended that a static pressure check be applied, if practicable, to ensure that the seal was installed correctly and wasn't damaged or defective.

B. SEAT SQUARENESS

To obtain the best performance from a mechanical face seal, the mating face must be held square to the axis of shaft rotation within recommended limits. An excessively out-of-square condition may cause increased leakage or shortened seal life. The following limits should normally be held if practicable.

Maximum Operational Speed in Revolutions per Minute	Maximum Total Indicator Reading Allowed in Inches
3,000	.0040
4,000	.0032
5,000	.0028
6,000	.0024
8,000	.0020
10,000	.0016
12,000	.0014
15,000	.0011
20,000	.0008
30,000	.0005
60,000	.0003
100,000	.0001

8
MECHANICAL FACE SEAL MALFUNCTIONS

A. SEAL MALFUNCTIONS AND PROBABLE CAUSES

Causes under this heading do not include improper design, materials, method of manufacture or quality assurance plan for application requirements.

1. Slight Initial Leakage Which Decreases

Should a seal start out with a slight leak, and if the leakage becomes progressively less with running, it indicates initial seal imperfections so small that the running-in process is sufficient to overcome them. Allow a reasonable time for the seal to run in; the process will take longer if the liquid being handled has good lubricating qualities.

2. Initial Leakage Which Continues

a. Secondary seals may have been damaged by sharp corners on bores, shafts, or sleeves.

b. Seal faces may have been distorted, cracked, scratched, chipped, or contaminated with dirt or other foreign matter before or during installation.

c. If the leakage is rather slight, the particular mechanism in which seal is installed may have excessive shaft-to-bore offset, shaft-to-bore misalignment, shaft run-out, shaft whip, or vibration.

3. Seal Emits Squeals or Chirps

Seal is running dry because of vacuum in the sealed system, air trapped in the sealed system, cavitation, or vaporization of liquid between sealing faces. The cause may be excessive speed or temperature, or insufficient system pressure.

4. Seal Spits and Sputters

Liquid film is vaporizing and flashing out of seal faces because of excessive speed or temperature, or because of incorrect pressure difference between inside and outside.

5. Seal Leaks and Ices

See preceding paragraph.

6. Black Powder Shows Up Outside Seal Faces

a. Liquid at dynamic sealing faces is insufficient for lubrication.

b. Liquid film between dynamic sealing faces is flashing or evaporating, leaving a residue which is grinding away the carbon or plastic seal face.

c. Pressure in the system being sealed is excessive for the seal or liquid being sealed.

7. Short Seal Life

a. Fluid being sealed is excessively abrasive, causing excessive wear.

b. Bearings or gears in mechanism being sealed have worn, allowing excessive whip, vibration, end-play, or frictional heat to develop.

B. RECORDING AND REPORTING SEAL MALFUNCTIONS

1. Before removing seal from mechanism remember that once the seal is disturbed in any manner, it cannot be operated again without both opposing faces being resurfaced or replaced.

2. If possible, photograph the seal area of the mechanism before removal of the seal.

3. Remove the malfunctioning seal with the greatest care possible. Rough handling of the seal, or carelessness at this point, will only make the analysis of failure more difficult or impossible.

4. If possible, photograph the seal immediately after removal.

5. Place seal in a protective container.

6. Carefully examine mechanism from which seal was removed, noting any unusual conditions, such as loose bearings or shaft, worn gears or cams.

7. Record the hours of operation or the number of mechanism cycles completed before the advent of leakage.

8. Record the amount and frequency of leakage; for example, "one drop every 90 seconds," "seal sputtered continuously," "squealed during spin cycle."

9. Record information about mechanism speed, temperature, pressure and operating mode when seal failed or started malfunctioning.

10. If practical, preserve a representative sampling of the fluid being sealed or the product of leakage, or both.

11. Record all circumstances and suspicions before details are forgotten.

9
OBTAINING ENGINEERING AND COST PROPOSALS

A. WHEN TO OBTAIN

1. New Designs

The time to start thinking about sealing requirements and working with seal manufacturers is during the beginning stage of a new product design.

It is often wise to choose two or three potential vendors, each of whom manufactures a wide range of sealing devices, such as radial shaft seals, packings and mechanical face seals. This will ensure that engineering and cost proposals are based on the best sealing device for your application, rather than on components produced by companies of limited capabilities, who can supply only one or two basic types of seals.

Be frank with potential seal suppliers regarding the chances of your new product reaching the marketplace, and the expected volume of production if it does. Also be candid when specifying the maximum leakage that can be tolerated in your application, and the operational life required of the seal. Don't overspecify—don't underspecify! Be receptive to new approaches with which you may not be familiar. However, if you have questions, no matter how petty they may seem, ask them . . . your potential vendors will be better able to help you if you do.

2. Redesigns

If you are redesigning your product for longer life, improved performance, greater reliability, easier manufacturing or servicing, or cost reductions, consider how these changes will or could affect the seals you are now using. As a precaution, consult your present seal suppliers about the changes you are contemplating. If any redesigns on their parts are required, this would also be an opportune time to evaluate what another supplier might have to offer.

3. Multiple Sources

Have you considered what might happen if a single supplier suddenly was unable to supply you with seals?

More than one assembly line has been shut down for shortage of a single-source critical component. Hedging against such possibilities by increasing the stock reserve may provide some protection, but often is more expensive than obtaining a second source. Your company may also enjoy other advantages of multiple sources, such as cost savings, better quality, improved deliveries and service.

B. HOW TO OBTAIN

An orderly way to exchange information between your engineering and purchasing departments and a seal manufacturer, is to use a vendor's data sheet specifically designed for seal applications. All of the information requested on the data sheet is pertinent in designing or specifying an optimum seal, and should be answered as completely and accurately as possible to ensure the most successful and economical solution to a sealing application.

Take potential suppliers into your confidence—they really don't want to manufacture anything but their specialty—seals. Reputable concerns will honor your confidence, and would not jeopardize their valuable reputation in the eyes of either your company or its competitors by divulging information which is proprietary.

APPLICATION AND ENGINEERING DATA SHEET FOR MECHANICAL FACE SEALS

CUSTOMER INFORMATION

NAME________________ TITLE________________

COMPANY______________________ DATE________

ADDRESS________________ CITY & STATE____________

PHONE NUMBER____________ SALESMAN______________

GENERAL INFORMATION

Seal application____________ Cust. Part No.____________

Seal is for: Price Study ☐ Preproduction Prototypes ☐
Production ☐ Replacement of Malfunction Seal ☐
Maintenance ☐ Other ☐

Est. quantity required__________ Mo. ☐ Yr. ☐
Approx. price________________________
Lead time before samples and/or production runs____________

QUALITY CONTROL INFORMATION

Customer will use: Statistical sampling ☐
100% inspection ☐ Spot check ☐ None ☐

Other____________________________
What characteristics are considered major______________
minor____________________________

If statistical sampling is used, what AQL is used for
Major defects____________ Minor defects____________
Cumulative total________________________
Remarks____________________________
(May be detailed under Special Customer Requirements)

INTERNATIONAL PACKINGS CORPORATION
BRISTOL, NEW HAMPSHIRE

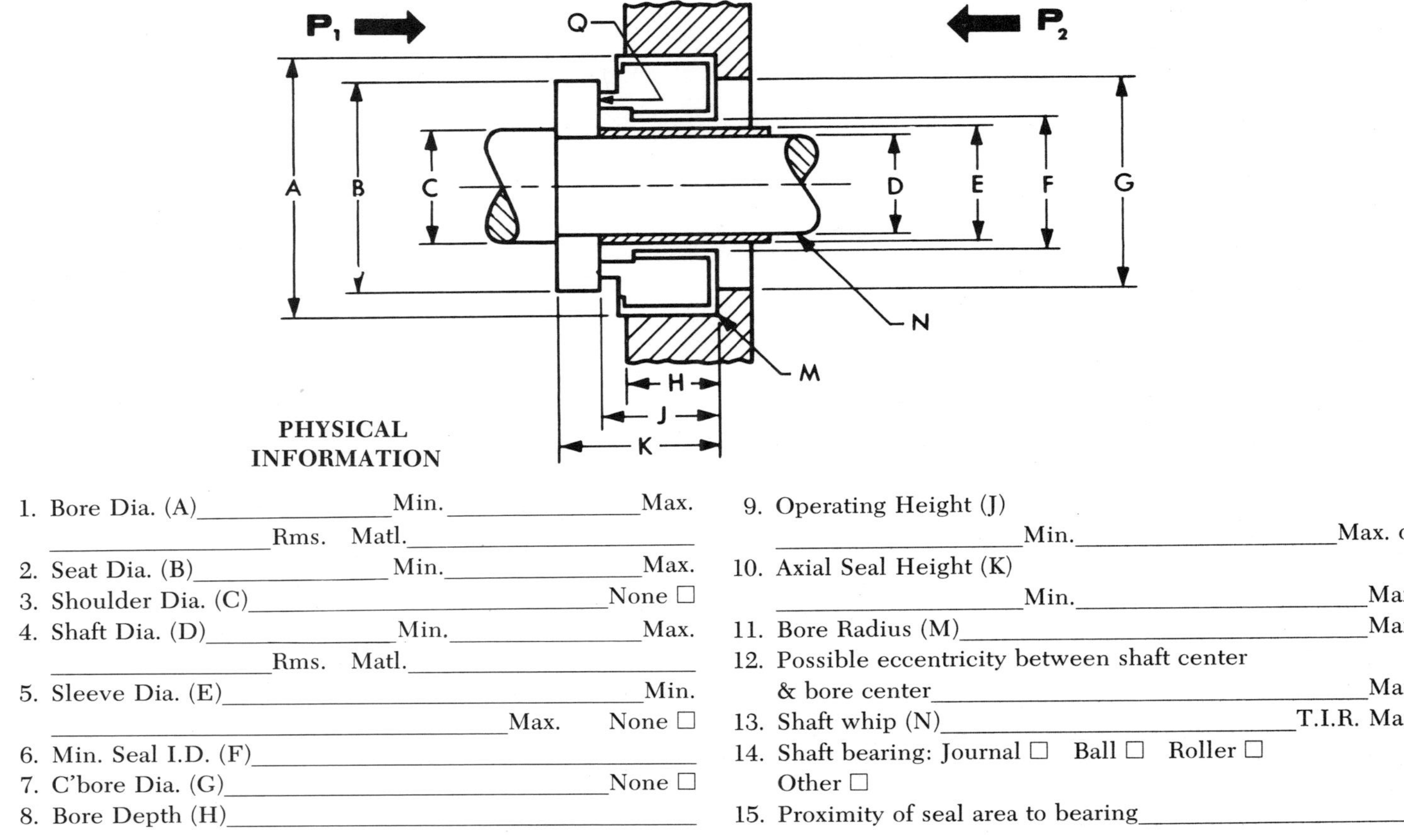

PHYSICAL INFORMATION

1. Bore Dia. (A)________Min. ________Max.
   ________Rms. Matl.________
2. Seat Dia. (B)________Min.________Max.
3. Shoulder Dia. (C)________None ☐
4. Shaft Dia. (D)________Min.________Max.
   ________Rms. Matl.________
5. Sleeve Dia. (E)________Min.
   ________Max. None ☐
6. Min. Seal I.D. (F)________
7. C'bore Dia. (G)________None ☐
8. Bore Depth (H)________
9. Operating Height (J)
   ________Min.________Max. or
10. Axial Seal Height (K)
    ________Min.________Max.
11. Bore Radius (M)________Max.
12. Possible eccentricity between shaft center
    & bore center________Max.
13. Shaft whip (N)________T.I.R. Max.
14. Shaft bearing: Journal ☐ Ball ☐ Roller ☐
    Other ☐
15. Proximity of seal area to bearing________

16. Shaft rotates ☐ Bore rotates ☐
    Normal__________Rpm. with
    Overspeed__________Rpm.__________% time
    or oscillates__________Degrees
    at__________Degrees/second
17. Out-of-square (Q)_____T.I.R.
18. Viewed from P_1 rotation is
    Clockwise ☐ Counterclockwise ☐
    Bidirectional ☐ Oscillates ☐

ENVIRONMENTAL INFORMATION

P_1

1. Medium
   Name__________
   Composition__________
   Specific gravity__________
   pH factor__________
   Nature & quantity of solids in medium__________
   Medium is ☐ Corrosive ☐ Abrasive
   ☐ Flammable ☐ Volatile
   Medium will ☐ Crystallize ☐ Polymerize
   ☐ Other__________
   Vapor pressure at operating temperature (if above atmospheric pressure)__________Psi.
2. Lubrication at P_1 ☐ Dry ☐ Mist ☐ Splash
   ☐ Immersion ☐ Intermittent; present __________% of time or (cycle)__________
3. Temperature__________°F. Min.
   __________°F. Max.__________°F. norm (oper)
4. Pressure (give units)
   Min.__________Max.__________
   Norm__________
   Additional information not covered above__________

P_2

1. Medium
   Name__________
   Composition__________
   Specific gravity__________
   pH factor__________
   Nature & quantity of solids in medium__________
   Medium is ☐ Corrosive ☐ Abrasive
   ☐ Flammable ☐ Volatile
   Medium will ☐ Crystallize ☐ Polymerize
   ☐ Other__________
   Vapor pressure at operating temperature (if above atmospheric pressure)__________Psi.

2. Lubrication at P_2 ☐ Dry ☐ Mist ☐ Splash
☐ Immersion ☐ Intermittent; present ____________ % of time or (cycle) ____________
3. Temperature ____________ °F. Min. ____________ °F. Max. ____________ °F. norm (oper)
4. Pressure (give units)
Min. ____________ Max. ____________
Norm ____________
Additional information not covered above ____________

GENERAL DATA

1. Max. leak rate desired:
____________ CC/Min. or ____________ CC/Hr.
2. Approximate life required (no safety factor)
____________ hours.
3. Describe any test procedure that seal must pass (supply specs if possible). ____________
4. Supply samples of mediums to be sealed if possible.
5. Supply prints of area surrounding seal if possible.
6. If this request is to supplement or replace a present source or type of seal, supply us with all possible data and/or history of seal now used.
7. Are samples necessary? ☐ Yes ☐ No

MARKING & PACKAGING INFORMATION

1. What identification must appear on seal? ____________
2. Must this identification appear on samples, if supplied?
☐ No ☐ Yes
3. What method(s) of marking may be used? ____________
4. Other requirements (style, size, depth, spacing, specs) ____________
5. Special packaging requirements ____________

SPECIAL CUSTOMER REQUIREMENTS

1. Materials and/or methods of construction ____________
2. Dimensional and/or load requirements ____________
3. Other ____________

10
MATERIALS OF CONSTRUCTION FOR VARIOUS FLUID SERVICES

A. GENERAL

The materials of construction for service with various fluids listed in this section represent a compilation of published recommendations of various mechanical face seal manufacturers. Such a listing must naturally be used with caution, since in many cases more suitable or economical materials can be used. Many factors must be considered when choosing materials for a specific application; these include the design of the seal, operating temperatures and speeds, surrounding materials, and the required operating life of the seal.

Many new materials are constantly being made available and evaluated for their possible use in mechanical face seals. The fact that today's mechanical face seals markedly outperform those of yester-year is in large part attributable to advances in materials technology.

B. CLARIFICATION OF SIMILAR TERMS

The term *fluorocarbon* refers to the more or less rigid fluorocarbon resins TFE and FEP, whereas the term *fluoroelastomer* refers to this class of elastomeric artificial rubber compounds. Carbon graphite refers to the family of dark and somewhat brittle carbonous bearing materials often used for one of the dynamic sealing surfaces in mechanical face seals.

TABLE V

Commonly Used Materials of Construction

Liquid	Lubricant?	Secondary Seals	Seal Washer	Metal Parts	Counterface	Spring(s)
Acetaldehyde	N	Fluorocarbon Carbon Graphite	Carbon Graphite	316 SS	Stellite Face Ceramic Stellite 316 SS	316 SS
Acetate Solvents	N	Fluorocarbon	Carbon Graphite	316 SS	Stellite Face Stellite	SS
Acetates, Esters, and Petroleum Derivatives	Y	Fluorocarbon	Carbon Graphite	316 SS	Stellite Face Stellite	SS
Acetic Anhydride	Y	Fluorocarbon	Carbon Graphite	316 SS	Ceramic Face Ceramic	316 SS
Acetone	N	Fluorocarbon Glass Filled-Fluorocarbon	Carbon Graphite	316 SS	Stellite Face Ceramic Bronze	316 SS
Acetone Cyanahydrin	N	Fluorocarbon	Carbon Graphite	SS	Ceramic	SS
Acetylene–Liquid	N	Nitrile*	Carbon Graphite	316 SS Steel	Ni-Resist Stellite Face	SS
Acetylene Gas with Water or Liquid H.C.	N	Fluorocarbon Nitrile*	Carbon Graphite	316 SS	Ceramic Ni-Resist	SS
Acetylene Tetrabromide	N	Fluorocarbon	Carbon Graphite	SS	Ceramic	SS
Acid, Acetic–Glacial	Y	Fluorocarbon Silicon	Carbon Graphite	316 SS	Ceramic Ceramic Face	316 SS

* Commonly specified by some seal manufacturers; however, more suitable materials may be available.

(Continued)

TABLE V (Cont'd)

Liquid	Lubricant?	Secondary Seals	Seal Washer	Metal Parts	Counterface	Spring(s)
Acid, Acetic—Petroleum Crude	Y	Fluorocarbon	Carbon Graphite	316 SS	Ceramic Stellite Face	316 SS
Acid, Acetic Propionic	Y	Fluorocarbon	Carbon Graphite	316 SS	Ceramic Face Stellite	SS
Acid, Acetic—Vapors	N	Fluorocarbon Silicon	Carbon Graphite	316 SS Carp. 20	Ceramic Ceramic Face	Carp. 20
Acid, Acetic—Wood Crude	Y	Fluorocarbon	Carbon Graphite	316 SS	Ceramic Ceramic Face	316 SS
Acid, Acetic—150 psi, 400°F	N	Fluorocarbon	Carbon Graphite	316 SS	Ceramic Stellite Face	316 SS
Acid, Alkyl-Arylsulfonic	Y	Fluorocarbon	Carbon Graphite	316 SS	Ceramic	316 SS
Acid, Arsenic	Y	Fluorocarbon Fluoroelastomer	Carbon Graphite	316 SS	Ceramic Face Ceramic	316 SS
Acid, Benzoic Sol.	Y	Fluorocarbon Glass Filled-Fluorocarbon	Carbon Graphite	316 SS	Stellite Face Ceramic	316 SS
Acid, Boric	Y	Fluorocarbon Chloroprene	Carbon Graphite	316 SS	Ceramic Ceramic Face	SS
Acid, Butyric	Y	Fluorocarbon Fluoroelastomer	Carbon Graphite	316 SS	Ceramic Ceramic Face	316 SS
Acid, Carbolic (Phenol)	Y	Fluorocarbon Fluoroelastomer	Carbon Graphite	316 SS	Stellite Stellite Face Ceramic	316 SS

Acid, Carbonic	Y	Fluorocarbon Chloroprene Nitrile	Carbon Graphite	316 SS	Stellite Ceramic Ceramic Face	SS
Acid, Chloroacetic – Mono	Y	Fluorocarbon	Fluorocarbon- Asbestos Carbon Graphite	316 SS Hast. "B"	Ceramic Ceramic Face	SS-Fluorocarbon Lined Hast. "B"
Acid, Chloroacetic – Di		Fluorocarbon	Fluorocarbon- Asbestos	316 SS	Ceramic	SS-Fluorocarbon Lined
Acid, Chloroacetic – Tri		Fluorocarbon	Fluorocarbon- Asbestos	316 SS	Ceramic	SS-Fluorocarbon Lined
Acid, Chlorosulfonic	Y	Fluorocarbon	Glass Filled- Fluorocarbon Fluorocarbon- Asbestos	Hast. "B" 316 SS	Ceramic Hast. "B" with Ceramic Face	Hastelloy SS-Fluorocarbon Lined
Acid, Chromic – 10%	Y	Fluorocarbon	Carbon Graphite	316 SS	Ceramic	316 SS
Acid, Chromic – 25%	Y	Fluorocarbon	Carbon Graphite	Carp. 20	Ceramic	Carp. 20
Acid, Citric	Y	Fluorocarbon Nitrile	Carbon Graphite	316 SS	Ceramic Ceramic Face	316 SS
Acid, Cresols (Cresylic)	Y	Fluorocarbon Fluoroelastomer	Carbon Graphite	316 SS	Ceramic Stellite Stellite Face	316 SS
Acid, Cresylic	See Acid, Cresols (Cresylic)					
Acid, Crotonic	Y	Fluorocarbon	Carbon Graphite	316 SS	Stellite Face Ceramic	SS
Acid, Fluosilicic		Fluorocarbon	Carbon Graphite	Monel	Carbon Graphite	Monel
Acid, Formic	Y	Fluorocarbon	Carbon Graphite	316 SS	Ceramic	316 SS

(Continued)

TABLE V (Cont'd)

Liquid	Lubricant?	Secondary Seals	Seal Washer	Metal Parts	Counterface	Spring(s)
Acid, Formic—60%, Boiling		Fluorocarbon	Fluorocarbon-Asbestos	316 SS	Ceramic	SS-Fluorocarbon Lined
Acid, Fumaric		Fluorocarbon	Carbon Graphite	316 SS	Ceramic	316 SS
Acid, Glutaric		Fluorocarbon	Carbon Graphite	316 SS	Ceramic	316 SS
Acid, Hydrobromic	Y	Fluorocarbon Fluoroelastomer	Carbon Graphite Fluorocarbon-Asbestos	Hast. "B" 316 SS	Ceramic	SS-Fluorocarbon Lined Hast. "B"
Acid, Hydrochloric or Muriatic	Y	Fluorocarbon	Carbon Graphite	Hast. "B"	Ceramic	Hastelloy
Acid, Hydrocyanic (Prussic Acid)	Y	Fluorocarbon Fluoroelastomer	Carbon Graphite	316 SS	Ceramic Ceramic Face	316 SS
Acid, Hydrofluoric		Fluorocarbon Fluoroelastomer	Carbon Graphite	316 SS Monel	Stellite	Monel
Acid, Hydrofluosilicic	See Acid, Fluosilicic					
Acid, Hydroformic	N	Fluorocarbon	Carbon Graphite	316 SS	Ceramic	316 SS
Acid, Hydroxyacetic	Y	Fluorocarbon	Carbon Graphite	316 SS Hast. "B"	Ceramic	316 SS Hast. "B"
Acid, Hypochlorous	Y	Fluorocarbon	Carbon Graphite	Hast. "C"	Ceramic	Hastelloy
Acid, Lactic	Y	Fluorocarbon	Carbon Graphite	316 SS	Ceramic Stellite Face	316 SS
Acid, Maleic	Y	Fluorocarbon Fluoroelastomer	Carbon Graphite	316 SS	Ceramic Face Ceramic	316 SS

Acid, Malic		Fluoroelastomer	Carbon Graphite	316 SS	Ceramic	316 SS
Acid, Mine – Non-Oxidizing	N	Fluorocarbon	Carbon Graphite	316 SS Monel	Ceramic	Monel
Acid, Mine – Oxidizing	N	Fluorocarbon Chloroprene	Carbon Graphite	316 SS	Ceramic	SS
Acid, Naphthenic	Y	Fluorocarbon	Carbon Graphite	316 SS	Stellite Face Ceramic	SS
Acid, Nitrating (mixed acid)	Y	Fluorocarbon	Carbon Graphite Fluorocarbon- Asbestos	316 SS	Stellite Ceramic	SS SS-Fluorocarbon Lined
Acid, Nitric <70%		Fluorocarbon Filled- Fluorocarbon Fluoroelastomer	Carbon Graphite Filled- Fluorocarbon	316 SS 304 SS	Ceramic Ceramic Face Titanium Stellite	316 SS 304 SS
Acid, Nitric >70%		Fluorocarbon Fluoroelastomer	Ceramic Ceramic Faced- Titanium Glass Filled- Fluorocarbon	304 SS	Ceramic Ceramic Faced- Titanium	304 SS
Acid, Nitric – Fuming		Fluorocarbon	Asbestos-Filled- Fluorocarbon	SS-Fluorocarbon Lined	Ceramic	SS-Fluorocarbon Lined
Acid, Oleic	Y	Fluorocarbon	Carbon Graphite	316 SS	Ceramic Face Ceramic	SS
Acid, Oxalic – Weak	Y	Fluorocarbon Chloroprene * Fluoroelastomer	Carbon Graphite	316 SS	Ceramic Ceramic Face	316 SS

* Commonly specified by some seal manufacturers; however, more suitable materials may be available.

(Continued)

TABLE V (Cont'd)

Liquid	Lubricant?	Secondary Seals	Seal Washer	Metal Parts	Counterface	Spring(s)
Acid, Oxalic – 50%, Boiling		Fluorocarbon	Carbon Graphite	316 SS	Ceramic	316 SS
Acid, Palmitic	Y	Fluorocarbon Fluoroelastomer Nitrile	Carbon Graphite	316 SS	Ceramic	316 SS
Acid, Phenylacetic	Y	Fluorocarbon	Carbon Graphite	316 SS	Ceramic Stellite Stellite Face	316 SS
Acid, Phosphoric <5%		Fluorocarbon Fluoroelastomer Nitrile	Carbon Graphite	316 SS	Ceramic	316 SS
Acid, Phosphoric – 5–45%		Fluorocarbon Fluoroelastomer	Carbon Graphite	316 SS	Ceramic Stellite Face	316 SS
Acid, Phosphoric >45%		Fluorocarbon	Carbon Graphite	Carp. 20	Ceramic Stellite Face	Carp. 20
Acid, Phthalic		Fluorocarbon	Carbon Graphite	316 SS	Ceramic	316 SS
Acid, Picric – Molten	N	Fluorocarbon	Carbon Graphite	316 SS	Ceramic Stellite Face	SS
Acid, Picric – Water Solution	Y	Fluorocarbon	Carbon Graphite	316 SS	Ceramic Face Ceramic	SS
Acid, Propionic	Y	Fluorocarbon Fluoroelastomer	Carbon Graphite	316 SS Monel	Ceramic Face Ceramic Stellite	316 SS
Acid, Pyroligneous (wood vinegar)	Y	Fluorocarbon	Carbon Graphite	316 SS	Ceramic	SS

Acid, Pyrosulfuric (Oleum)		Fluorocarbon Fluoroelastomer	Carbon Graphite Fluorocarbon-Filled Ceramic	Carp. 20 316 SS	Ceramic	Carp. 20 316 SS
Acid, Sludge	N	Fluorocarbon	Carbon Graphite	Hast. "C"	Ceramic	Hast. "C"
Acid, Stearic	N	Fluorocarbon	Carbon Graphite	316 SS	Ceramic Face Ceramic	316 SS
Acid, Stearic – 200°F	Y	Fluorocarbon	Carbon Graphite	316 SS	Ceramic	SS
Acid, Stearic & Oleic	N	Fluorocarbon	Carbon Graphite	316 SS	Ceramic Ceramic Face	316 SS
Acid, Succinic		Fluorocarbon	Carbon Graphite	316 SS	Ceramic	316 SS
Acid, Sulfuric up to 95%		Fluorocarbon Fluoroelastomer	Carbon Graphite	Carp. 20 316 SS	Ceramic Ceramic Face	316 SS Carp. 20
Acid, Sulfurous	Y	Fluorocarbon	Carbon Graphite	Hast. "C"	Ceramic Ceramic Face	Hast. "C"
Acid, Tannic	Y	Fluorocarbon Chloroprene	Carbon Graphite	316 SS	Stellite Stellite Face	SS
Acid, Tartaric	Y	Fluorocarbon Chloroprene *	Carbon Graphite	316 SS	Stellite Stellite Face	SS
Acid, Terephthalic	Y	Fluorocarbon	Carbon Graphite	316 SS	Ceramic	SS
Acid, Terephthalic, Water Slurry, 10% Nitric Acid	N	Fluorocarbon	Carbon Graphite	316 SS	Ceramic	SS
Acrolein	Y	Fluorocarbon	Carbon Graphite	316 SS	Ceramic	316 SS

* Commonly specified by some seal manufacturers; however, more suitable materials may be available.

(Continued)

TABLE V (Cont'd)

Liquid	Lubricant?	Secondary Seals	Seal Washer	Metal Parts	Counterface	Spring(s)
Acrylonitrile	Y	Fluorocarbon	Carbon Graphite	316 SS	Stellite Stellite Face Ceramic 316 SS Ni-Resist	316 SS
Alcohol		Nitrile	Carbon Graphite	316 SS	Bronze	316 SS
Alcohol, Amyl	Y	Fluorocarbon	Carbon Graphite	316 SS Brass	Ceramic Stellite Face	SS
Alcohol, Butyl	Y	Nitrile Fluoroelastomer	Carbon Graphite	316 SS Brass	Ceramic Stellite Face Bronze	316 SS
Alcohol, Diacetone	See Diacetone					
Alcohol, Ethyl (Ethanol)	Y	Nitrile	Carbon Graphite	316 SS Brass	Stellite Ceramic Stellite Face Bronze	316 SS
Alcohol, Grain	See Alcohol, Ethyl					
Alcohol, Isobutyl (Isobutanol)	Y	Fluoroelastomer Fluorocarbon Nitrile *	Carbon Graphite	316 SS	Ni-Resist Stellite Stellite Face	316 SS
Alcohol, Isopropyl (Isopropanol)	Y	Nitrile Fluoroelastomer	Carbon Graphite	316 SS Steel	Ni-Resist Stellite Face	316 SS
Alcohol & Lub. Oil	Y	Nitrile	Carbon Graphite	316 SS Steel	Ni-Resist Stellite Face	SS

Alcohol, Methyl (Methanol)	Y	Nitrile	Carbon Graphite	316 SS Brass	Ceramic Stellite Face Bronze	316 SS
Alcohol, Octyl (Octanol)	Y	Fluorocarbon Nitrile	Carbon Graphite	316 SS	Ceramic Stellite Face	SS
Alcohol, Propyl (Propanol)		Fluoroelastomer	Carbon Graphite	316 SS	Ni-Resist	316 SS
Alcohol, Sulfonated Fatty	Y	Fluorocarbon	Carbon Graphite	Hast. "C" SS	Stellite Ceramic	SS
Aldehyde	Y	Fluorocarbon	Carbon Graphite	316 SS	Ceramic Stellite	SS
Alkali Cleaner	Y	Fluorocarbon Chloroprene	Carbon Graphite	316 SS	Ceramic	SS
Alkanes	Y	Nitrile	Carbon Graphite	316 SS Steel	Ni-Resist Stellite Face	SS
Alkyd Resins	Y	Fluorocarbon	Carbon Graphite	316 SS	Ceramic Stellite Face	SS
Alkylate, Light	Y	Fluorocarbon Fluoroelastomer	Carbon Graphite	316 SS	Stellite Face Stellite	SS
Alkyl Benzene	Y	Fluorocarbon Fluoroelastomer	Carbon Graphite	316 SS	Ceramic Stellite Face	SS
Allyl Acetone	Y	Fluorocarbon	Carbon Graphite	316 SS	Ceramic Stellite Face	SS
Allyl Amine	N	Fluorocarbon	Carbon Graphite	316 SS	Ceramic Stellite Face	SS
Allyl Chloride	N	Fluorocarbon	Carbon Graphite	316 SS	Ceramic Stellite Face	316 SS

* Commonly specified by some seal manufacturers; however, more suitable materials may be available.

(Continued)

TABLE V (Cont'd)

Liquid	Lubricant?	Secondary Seals	Seal Washer	Metal Parts	Counterface	Spring(s)
Alpha Picoline	N	Fluorocarbon	Carbon Graphite	316 SS	Stellite Ceramic	SS
Alum Solution	N	Fluorocarbon Nitrile	Carbon Graphite	Carp. 20	Stellite Ceramic	SS
Aluminum Chloride	N	Fluorocarbon Nitrile	Carbon Graphite	Hast. "B"	Ceramic	Hast. "B"
Aluminum Hydroxide	N	Nitrile	Carbon Graphite	316 SS Steel	Ni-Resist Stellite Face	SS
Aluminum Sulfate	N	Fluorocarbon Nitrile	Carbon Graphite	Carp. 20	Ceramic Ceramic Face	SS
Amine, Diethanol	Y	Fluorocarbon	Carbon Graphite	316 SS	Stellite Stellite Face	SS
Amine, Fat Condensate	Y	Fluorocarbon	Carbon Graphite	316 SS	Stellite Face Ceramic	SS
Ammonia, Anhydrous		Chloroprene Nitrile	Carbon Graphite	316 SS Steel	Stellite Cast Iron Tungsten Carbide Stellite Face Ni-Resist	316 SS
Ammonia, Aqueous	Y	Chloroprene	Carbon Graphite	316 SS Steel	Cast Iron Tungsten Carbide	SS
Ammonia, Liquors & Oil	Y	Chloroprene Fluorocarbon	Carbon Graphite	316 SS Steel	Cast Iron Stellite Face	SS
Ammonia and Oil		Nitrile	Carbon Graphite	316 SS	316 SS	316 SS

Ammonium Bicarbonate	N	Chloroprene	Carbon Graphite	316 SS Steel	Cast Iron Stellite Face	SS
Ammonium Bifluoride	N	Fluorocarbon	Carbon Graphite	Monel	Stellite Stellite Face	Monel
Ammonium Carbonate	N	Chloroprene	Carbon Graphite	316 SS Steel	Cast Iron Stellite Face	SS
Ammonium Chloride (Sal Amoniac)	N	Fluorocarbon Chloroprene	Carbon Graphite	Hast. "B"	Ceramic Ceramic Face	Hast. "B"
Ammonium Hydroxide	Y	Chloroprene Nitrile *	Carbon Graphite	316 SS Steel	Cast Iron Stellite Face Stellite	316 SS
Ammonium Nitrate Sol. (40–80%) (10–140°F)	N	Fluorocarbon Chloroprene *	Glass Filled-Fluorocarbon Carbon Graphite	SS Carp. 20	Ceramic Ceramic Face	SS Carp. 20
Ammonium Phosphate – Mono Basic	Y	Fluorocarbon Chloroprene	Carbon Graphite	316 SS	Ceramic Ceramic Face	SS
Ammonium Phosphate – Di Basic	Y	Fluorocarbon Chloroprene	Carbon Graphite	316 SS	Ceramic Ceramic Face	SS
Ammonium Phosphate – Tri Basic	Y	Fluorocarbon Chloroprene	Carbon Graphite	316 SS	Ceramic Ceramic Face	SS
Ammonium Sulfate – Heavy	Y	Nitrile	Carbon Graphite	316 SS Monel	Ceramic	Monel
Ammonium Sulfate – Weak	Y	Nitrile	Carbon Graphite	316 SS Monel	Ceramic	Monel
Ammonium Thiocyanate	Y	Fluorocarbon	Carbon Graphite	316 SS	Ceramic Stellite Face	SS

* Commonly specified by some seal manufacturers; however, more suitable materials may be available.

(Continued)

TABLE V (Cont'd)

Liquid	Lubri-cant?	Secondary Seals	Seal Washer	Metal Parts	Counterface	Spring(s)
Ammonium Thiocyanide	Y	Fluorocarbon	Carbon Graphite	316 SS	Ceramic Stellite Face	SS
Amyl Acetate	Y	Fluorocarbon EPR	Carbon Graphite	316 SS	Stellite Stellite Face Bronze	316 SS
Amyl Butyrate (Isoamyl butyrate)		Fluorocarbon	Carbon Graphite	316 SS	Bronze	316 SS
Amyl Formate		Fluorocarbon	Carbon Graphite	316 SS	Bronze	316 SS
Amyl Nitrate	Y	Fluorocarbon	Carbon Graphite	316 SS	Ceramic Stellite Face Bronze	316 SS
Amyl Propionic		Fluorocarbon	Carbon Graphite	316 SS	Bronze	316 SS
Aniline		Fluorocarbon	Carbon Graphite	316 SS	Bronze	316 SS
Aniline Dyes	Y	Fluorocarbon	Carbon Graphite	316 SS	Stellite Face Stellite	SS
Aniline Hydrochloride	N	Fluorocarbon	Carbon Graphite	Hast. "C"	Ceramic	Hast. "C"
Aniline – Liquid	Y	Fluorocarbon	Carbon Graphite	316 SS	Stellite Stellite Face	SS
Aniline – Vapors	N	Fluorocarbon	Carbon Graphite	316 SS	Stellite Stellite Face	SS
Antibiotic Fermentation Broth – No Solvent	N	Fluorocarbon Chloroprene	Carbon Graphite	316 SS	Ceramic Ceramic Face	SS

Antibiotic Fermentation Broth—With Solvent	N	Fluorocarbon	Carbon Graphite	316 SS	Ceramic Ceramic Face	SS
Anti-Freeze, Water & Alcohols or Glycols	Y	Nitrile	Carbon Graphite	316 SS	Ceramic Face Ni-Resist	SS
Anti-Freeze	See Ethylene Glycol					
Argon		Nitrile	Carbon Graphite	316 SS	316 SS	316 SS
Arochlor 1248	Y	Fluorocarbon	Carbon Graphite	316 SS	Ni-Resist Cast Iron	SS
Aromatic Fuels—<20% Aromatics		Nitrile Fluorocarbon Fluoroelastomer	Carbon Graphite	316 SS	Bronze	316 SS
Aromatic Fuels—>20% Aromatics		Fluorocarbon	Carbon Graphite	316 SS	Bronze Ceramic Ceramic Face	316 SS
Asphalt, Emulsified		Fluorocarbon Fluoroelastomer	Carbon Graphite	316 SS	Stellite Face	316 SS
Barium Chloride Sol.—0–20%	N	Nitrile	Carbon Graphite	316 SS	Ceramic	316 SS
Barium Hydroxide	N	Fluorocarbon Nitrile	Carbon Graphite	316 SS	Stellite Face	316 SS
Barium Nitrate Sol.	N	Fluorocarbon Fluoroelastomer	Carbon Graphite	316 SS	Ceramic	SS
Barium Sulfide	N	Fluorocarbon	Carbon Graphite	316 SS	Ceramic Ceramic Face	SS
Bauxite	N	Fluorocarbon	Carbon Graphite	316 SS	Stellite Ceramic Face	SS

(Continued)

TABLE V (Cont'd)

Liquid	Lubricant?	Secondary Seals	Seal Washer	Metal Parts	Counterface	Spring(s)
Beer	N	Nitrile	Carbon Graphite	316 SS	Stellite Face Ni-Resist	SS
Beer Wort to 212°F	N	Fluorocarbon Nitrile	Carbon Graphite	316 SS	Ceramic Ceramic Face	SS
Beet Juice and Pulp	N	Nitrile	Carbon Graphite	316 SS Brass	Ceramic Ceramic Face	SS
Beet Sugar Liquors	N	Nitrile Fluorocarbon	Carbon Graphite	316 SS Brass	Ceramic Ceramic Face	316 SS
Benzene (Aliphatic)		Nitrile	Carbon Graphite	316 SS	Ni-Resist	316 SS
Benzene (Benzol)	Y	Fluorocarbon Fluoroelastomer	Carbon Graphite	316 SS	Ceramic Stellite Face Bronze	316 SS
Benzene & Gas Mixtures—40%	Y	Fluorocarbon	Carbon Graphite	SS	Ceramic	SS
Benzine (Ligroin)	Y	Fluorocarbon Nitrile Fluoroelastomer	Carbon Graphite	316 SS	Ceramic Ni-Resist Stellite Face	316 SS
Benzol See Benzene (Benzol)						
Bichloride of Mercury	N	Fluorocarbon	Carbon Graphite	316 SS	Ceramic	316 SS
Black Sulfate Liquor	N	Fluorocarbon Fluoroelastomer	Carbon Graphite Ni-Resist	316 SS	Ceramic Stellite Face	SS
Bleach Solution	N	Fluorocarbon	Carbon Graphite	Hast. "C"	Ceramic	Hast. "C"

Bonderite Solution	N	Fluorocarbon Nitrile	Carbon Graphite	316 SS	Ceramic	SS
Borax Solution	N	Fluorocarbon Nitrile *	Carbon Graphite	316 SS	Ceramic Ceramic Face	SS
Bromine		Fluorocarbon	Fluorocarbon- Asbestos	316 SS	Ceramic	SS-Fluorocarbon Lined
Bromine, Dry	N	Fluorocarbon	Glass Filled- Fluorocarbon	Hastelloy	Ceramic	Hastelloy
Butadiene	Y	Fluorocarbon Nitrile *	Carbon Graphite	316 SS	Stellite Stellite Face Ni-Resist Tungsten Carbide	316 SS
Butane	Y	Nitrile Chloroprene	Carbon Graphite	316 SS Steel	Stellite Ni-Resist Stellite Face Tungsten Carbide	316 SS
Butanol See Alcohol, Butyl						
Butyl Acetate	N	Fluorocarbon	Carbon Graphite	316 SS	Ceramic Stellite Face Bronze	316 SS
Butylamine	Y	Fluorocarbon Nitrile *	Carbon Graphite	316 SS	Stellite Stellite Face 316 SS	316 SS
Butyl Butyrate		Fluorocarbon	Carbon Graphite	316 SS	Bronze	316 SS
Butylene	Y	Nitrile *	Carbon Graphite	316 SS Steel	Stellite Ni-Resist Tungsten Carbide	316 SS

* Commonly specified by some seal manufacturers; however, more suitable materials may be available.

(Continued)

TABLE V (Cont'd)

Liquid	Lubricant?	Secondary Seals	Seal Washer	Metal Parts	Counterface	Spring(s)
Butyl Formate		Fluorocarbon	Carbon Graphite	316 SS	Bronze	316 SS
Butyl Propionate		Fluorocarbon	Carbon Graphite	316 SS	Bronze	316 SS
Calcium Bisulfide (Calcium Hydrosulfide)	N	Fluorocarbon	Carbon Graphite	316 SS	Ceramic Ceramic Face	SS
Calcium Bisulfite	N	Fluorocarbon Fluoroelastomer	Carbon Graphite	316 SS	Ceramic Ceramic Face	SS
Calcium Carbonate Slurry	N	Nitrile Fluorocarbon	Carbon Graphite	316 SS Brass	Ceramic Ceramic Face	SS
Calcium Chloride Brine—Bronze Pump	N	Nitrile	Carbon Graphite	316 SS Brass	Ceramic Ceramic Face	SS
Calcium Chloride Brine—Iron Pump	N	Nitrile	Carbon Graphite	316 SS Steel	Ceramic Ni-Resist Ceramic Face	SS
Calcium Chloride Brine—Inhibited	N	Fluorocarbon Nitrile	Carbon Graphite	316 SS	Ceramic	SS
Calcium Hydroxide	N	Nitrile	Carbon Graphite	Carp. 20 Monel	Stellite Face	Carp. 20
Calcium Hypochlorite	N	Fluorocarbon Fluoroelastomer	Carbon Graphite	Hast. "B"	Ceramic Ceramic Face	Hast. "B"
Calcium Phosphate Slurry	N	Fluorocarbon Nitrile	Carbon Graphite	316 SS Brass	Ceramic	SS
Calgon (Tri Sodium Phosphate)	N	Nitrile	Carbon Graphite	316 SS Steel	Ni-Resist Ceramic Face	SS

Caliche Liquors	N	Fluorocarbon Fluoroelastomer	Carbon Graphite Ni-Resist	316 SS	Stellite Stellite Face	SS
Candelilla Wax	Y	Fluorocarbon Nitrile	Carbon Graphite	316 SS Steel	Ni-Resist Stellite Face	SS
Cane Sugar Liquors & Cane Juice	N	Nitrile	Carbon Graphite	316 SS Brass	Ceramic Ceramic Face	SS
Carbinol (Methanol)	See Alcohol, Methyl					
Carbon Bisulfide (Carbon Disulfide)	N	Fluorocarbon Fluoroelastomer	Carbon Graphite	316 SS	Ceramic Stellite Face Bronze	316 SS
Carbon Dioxide		Nitrile	Carbon Graphite	316 SS	Ni-Resist	316 SS
Carbon Disulfide	See Carbon Bisulfide					
Carbon Tetrachloride–Anhydrous	Y	Fluoroelastomer	Carbon Graphite	316 SS	Ceramic Bronze	316 SS
Carbon Tetrachloride–Wet	Y	Fluorocarbon	Carbon Graphite	Monel	Ceramic Stellite	Monel Hastelloy
Casein	Y	Fluorocarbon Nitrile	Carbon Graphite	316 SS	Ceramic Stellite Face	SS
Caustic Soda–10%, 200°F (Sodium Hydroxide)		Nitrile	Carbon Graphite	Monel	Stellite	Monel
Cellosolve		Fluorocarbon	Carbon Graphite	316 SS	Bronze	316 SS
Chloroacetaldehyde	Y	Fluorocarbon	Carbon Graphite	316 SS	Stellite Face Stellite	SS
Chlorex	Y	Fluorocarbon	Carbon Graphite	Monel Hast. "B"	Ceramic	Monel

(Continued)

TABLE V (Cont'd)

Liquid	Lubricant?	Secondary Seals	Seal Washer	Metal Parts	Counterface	Spring(s)
Chlorine – Wet		Fluorocarbon	Fluorocarbon-Asbestos Carbon Graphite	316 SS Hast. "C"	Ceramic	SS-Fluorocarbon Lined Hastelloy
Chlorine – Anhydrous Liquid	N	Fluorocarbon	Carbon Graphite	Hastelloy 316 SS	Ceramic Glass Filled-Fluorocarbon	Hastelloy
Chlorine – Dry	N	Fluorocarbon Fluoroelastomer	Carbon Graphite	Hastelloy 316 SS	Ceramic Ceramic Face	Hastelloy
Chlorobenzene – Mono	N	Fluorocarbon Fluoroelastomer	Carbon Graphite	316 SS	Bronze Stellite Stellite Face	316 SS
Chlorobenzene – Di	Y	Fluorocarbon Fluoroelastomer	Carbon Graphite	316 SS	Stellite Stellite Face	SS
Chlorobenzene – Tri	Y	Fluorocarbon Fluoroelastomer	Carbon Graphite	316 SS	Stellite Stellite Face	SS
Chloroethane – Di	See Ethylene Dichloride					
Chloroethane – Tri, Dry	N	Fluorocarbon	Carbon Graphite	316 SS	Ceramic Ni-Resist Stellite Face	SS
Chloroethane – Tri, Wet	Y	Fluorocarbon	Carbon Graphite	316 SS Monel	Ceramic Stellite Face	Monel
Chloroethylene – Tri, Dry	N	Fluorocarbon Fluoroelastomer	Carbon Graphite	316 SS	Stellite Ceramic Stellite Face	SS

Chloroethylene – Tri, Wet	Y	Fluorocarbon Fluoroelastomer	Carbon Graphite	Monel	Ceramic	Monel
Chloroform	Y	Fluorocarbon Fluoroelastomer	Carbon Graphite	316 SS	Stellite Face Ceramic Bronze	316 SS
Chlorohydrin – Di, 30%	N	Fluorocarbon	Carbon Graphite	316 SS	Stellite Ceramic	SS
Chlorohydrin – Tri, 10%	N	Fluorocarbon	Carbon Graphite	316 SS	Ceramic	SS
Chloromethyl Ether	Y	Fluorocarbon	Carbon Graphite	316 SS	Stellite Face Stellite	SS
Chrome Alum (Chromium Potassium Sulfate)	N	Fluorocarbon	Carbon Graphite	316 SS Monel	Ceramic	Monel
Clay (Slurry)	N	Nitrile	Carbon Graphite	316 SS	Ni-Resist Ceramic Ceramic Face	SS
Coal Tar	N	Fluorocarbon Glass Filled-Fluorocarbon	Carbon Graphite	316 SS	Stellite Ceramic Face Stellite Face	SS
Cocoanut – Fatty Acid	Y	Fluorocarbon	Carbon Graphite	316 SS	Stellite Stellite Face	SS
Cocoa Butter	Y	Nitrile	Carbon Graphite	316 SS Brass	Ceramic Ceramic Face	SS
Coffee Extract	N	Fluorocarbon	Carbon Graphite	316 SS	Stellite Ceramic	SS
Coke Breeze	N	Fluorocarbon	Carbon Graphite	316 SS	Ceramic	SS
Copper Acetate	N	Fluorocarbon EPR	Carbon Graphite	316 SS	Ceramic Ceramic Face	SS

(Continued)

TABLE V (Cont'd)

Liquid	Lubricant?	Secondary Seals	Seal Washer	Metal Parts	Counterface	Spring(s)
Copper Ammonium Acetate	N	Fluorocarbon Chloroprene	Carbon Graphite	316 SS	Ceramic Stellite Face	SS
Copperas (Green) (Ferrous Sulfate)	N	Fluorocarbon Fluoroelastomer	Carbon Graphite Ni-Resist	Hastelloy 316 SS	Ceramic Stellite Face	Hastelloy
Copper Chloride (Cupric) 10%—Room Temp.	N	Fluorocarbon Nitrile	Carbon Graphite	Hast. "C"	Ceramic	Hastelloy
Copper Chloride Brine—up to 75°F	N	Fluorocarbon Nitrile	Carbon Graphite	316 SS Hastelloy	Ceramic	Hastelloy
Copper Cyanide	N	Fluorocarbon Nitrile	Carbon Graphite	SS Carp. 20	Ceramic Stellite Face	SS
Copper Nitrate	N	Fluorocarbon Nitrile	Carbon Graphite	316 SS	Ceramic	SS
Copper Sulfate—Blue Vitriol	N	Fluorocarbon Nitrile	Carbon Graphite	316 SS	Ceramic	SS
Core Oils	Y	Nitrile	Carbon Graphite	316 SS	Stellite Stellite Face	SS
Corn Syrup	Y	Fluorocarbon Fluoroelastomer	Carbon Graphite	316 SS	Ceramic Stellite Face	SS
Creosote	Y	Fluoroelastomer Fluorocarbon	Carbon Graphite	316 SS	Tungsten Carbide Ceramic Stellite Face	316 SS
Cresol—Meta	Y	Fluorocarbon	Carbon Graphite	316 SS	Stellite Stellite Face	SS

Crotonaldehyde	Y	Fluorocarbon	Carbon Graphite	316 SS	Ceramic Stellite Face	316 SS
Cumene	N	Fluorocarbon	Carbon Graphite	316 SS	Ceramic Stellite Face	SS
Cuprous Ammonia Acetate	Y	Fluorocarbon Chloroprene	Carbon Graphite	316 SS	Ceramic Ni-Resist	SS
Cyclohexane	Y	Fluorocarbon Nitrile	Carbon Graphite	316 SS	Ceramic Stellite Face Bronze	316 SS
DDT Solutions (Kerosene Solvent)	Y	Nitrile Fluoroelastomer	Carbon Graphite	316 SS	Stellite Face Stellite	SS
DDT Solutions (Toluene Solvent)	Y	Fluorocarbon Fluoroelastomer	Carbon Graphite	316 SS	Ceramic Stellite Face	SS
De-Butanizer Reflux	Y	Nitrile	Carbon Graphite	316 SS Steel	Stellite Stellite Face	SS
De-Ethanizer Charge	Y	Nitrile	Carbon Graphite	316 SS Steel	Stellite Face Stellite	SS
De-Propanizer Reflux	Y	Nitrile	Carbon Graphite	316 SS Steel	Stellite Stellite Face	SS
Diacetone (Diacetone Alcohol)	Y	Fluorocarbon	Carbon Graphite	316 SS	Ceramic Bronze Stellite Face	316 SS
Diatomaceous Earth/H_2O (Diatomite/H_2O)	N	Nitrile	Carbon Graphite	316 SS Brass	Ceramic Ceramic Face	SS
Dibutyl "Cellosolve" Adipate	N	Fluorocarbon	Carbon Graphite	316 SS	Ceramic Stellite Face	SS
Diesel Fuel	See Oil, Diesel					

(Continued)

TABLE V (Cont'd)

Liquid	Lubricant?	Secondary Seals	Seal Washer	Metal Parts	Counterface	Spring(s)
Diethyl Carbonate (Ethyl Carbonate)	Y	Fluorocarbon Nitrile	Carbon Graphite	316 SS	Ceramic Stellite Face Bronze	316 SS
Diethylene Glycol	Y	Fluorocarbon Nitrile	Carbon Graphite	316 SS Steel	Stellite Face Bronze Ni-Resist	316 SS
Diethylene Triamine	N	Fluorocarbon	Carbon Graphite	316 SS	Stellite Face Ceramic	SS
Diethyl Maleate	Y	Fluorocarbon	Carbon Graphite	316 SS	Stellite Face Stellite	SS
Diisobutyl Ketone	Y	Fluorocarbon	Carbon Graphite	316 SS	Ceramic Stellite Face	SS
Dimethyl Formaldehyde	N	Fluorocarbon	Carbon Graphite	316 SS	Stellite Face Stellite	SS
Dimethyl Hydrazine	N	Fluorocarbon	Carbon Graphite	316 SS	Ceramic Stellite Face	SS
Dinitrochlorobenzene and Styrene	N	Fluorocarbon	Carbon Graphite	316 SS	Ceramic Stellite Face	SS
Dioctyl – Amine (Diethylhexylamine)	N	Fluorocarbon	Carbon Graphite	316 SS	Stellite Face Stellite	SS
Dioctyl – Phthalate (Diethylhexyl phthalate)	N	Fluorocarbon	Carbon Graphite	316 SS	Ceramic Stellite Face	SS
Diphenyl	N	Fluorocarbon Glass Filled-Fluorocarbon	Carbon Graphite	316 SS	Stellite Stellite Face	SS

Diphenyl Amine – 4%	N	Fluorocarbon Glass Filled-Fluorocarbon	Carbon Graphite	316 SS	Stellite Ceramic Face	SS
Diphenyl, Chlorinated	N	Fluorocarbon Glass Filled-Fluorocarbon	Carbon Graphite	316 SS	Stellite Stellite Face	SS
DNCB (Dinitro-Chloro-Benzene) (Chloro-Dinitrobenzene)	Y	Fluorocarbon	Carbon Graphite	316 SS	Ceramic	SS
Doctor Solution See Sodium Plumbite						
Dowtherm "A"	Y	Filled-Fluorocarbon	Carbon Graphite	316 SS	Stellite	316 SS
Dye Liquors	Y	Fluorocarbon	Carbon Graphite	316 SS	Ceramic	SS
Epichlorohydrin – 60%	N	Fluorocarbon	Carbon Graphite	316 SS Carp. 20	Ceramic	SS
Ester Type Plasticizer	Y	Fluorocarbon	Carbon Graphite	316 SS	Stellite Stellite Face	SS
Ethane	Y	Nitrile	Carbon Graphite	316 SS Steel	Cast Iron Tungsten Carbide	316 SS Steel
Ethanol See Alcohol, Ethyl						
Ethanolamine, Mono – With Copper	Y	Fluorocarbon	Carbon Graphite Glass Filled-Fluorocarbon	316 SS	Ceramic Stellite Face	316 SS
Ethanolamine, Mono – Without Copper	Y	Fluorocarbon	Carbon Graphite	316 SS	Ceramic Stellite Face	SS
Ethanolamine, Di-	Y	Fluorocarbon	Carbon Graphite	316 SS	Stellite Face Stellite	SS

(Continued)

TABLE V (Cont'd)

Liquid	Lubricant?	Secondary Seals	Seal Washer	Metal Parts	Counterface	Spring(s)
Ethanolamine, Tri-	Y	Fluorocarbon	Carbon Graphite	316 SS	Ceramic Stellite Face	SS
Ether	Y	Fluorocarbon	Carbon Graphite	316 SS	Ceramic Stellite Face Bronze	316 SS
Ethyl Acetate	Y	Fluorocarbon	Carbon Graphite	316 SS	Ceramic Stellite Face Bronze	316 SS
Ethyl Benzoate		Fluorocarbon	Carbon Graphite	SS	Ceramic	SS
Ethyl Bromide	N	Fluorocarbon	Carbon Graphite	SS Carp. 20	Ceramic Ceramic Face	SS
Ethyl Butyrate		Fluorocarbon	Carbon Graphite	316 SS	Bronze	316 SS
Ethyl Cellulose	N	Fluorocarbon	Carbon Graphite	316 SS	Ceramic Stellite Face	SS
Ethyl Chloride	N	Fluorocarbon Fluoroelastomer	Carbon Graphite	Monel Carp. 20	Ceramic Ceramic Face	Monel Hastelloy
Ethyl Dichloride		Fluorocarbon Fluoroelastomer	Carbon Graphite	Carp. 20	Ceramic Face Ceramic	Carp. 20
Ethylene	Y	Fluorocarbon Nitrile Glass Filled- Fluorocarbon	Carbon Graphite	316 SS	Tungsten Carbide	316 SS
Ethylene Bromide (Ethylene Dibromide)	Y	Fluorocarbon	Carbon Graphite	316 SS Carp. 20	Ceramic Ceramic Face	316 SS

Ethylene Chloride (Ethylene Dichloride)	See Ethylene Dichloride					
Ethylene Dichloride	Y	Fluorocarbon Fluoroelastomer	Carbon Graphite	316 SS Hastelloy Monel	Stellite Face Ceramic	Hastelloy
Ethylene Glycol	Y	Fluorocarbon Nitrile	Carbon Graphite	316 SS Steel	Ceramic Stellite Face Ni-Resist Stellite Bronze	316 SS
Ethylene Oxide	N	Fluorocarbon Nitrile	Carbon Graphite	316 SS	Stellite Stellite Face Bronze	316 SS
Ethylene Thiourea	N	Fluorocarbon	Carbon Graphite	316 SS	Stellite Stellite Face	SS
Ethyl Formate		Fluorocarbon	Carbon Graphite	316 SS	Bronze	316 SS
Ethyl Propionate		Fluorocarbon	Carbon Graphite	316 SS	Bronze	316 SS
Ethyl Pyridine	N	Fluorocarbon	Carbon Graphite	316 SS	Stellite Face Stellite	SS
Ethyl Sulfate (Diethyl Sulfate)	N	Fluorocarbon	Carbon Graphite	316 SS	Ceramic	SS
Fatty Acids	Y	Fluorocarbon	Carbon Graphite	316 SS	Ceramic Stellite Face	SS
Ferric Chloride—40%	N	Fluorocarbon	Carbon Graphite	Hastelloy	Ceramic	Hastelloy
Ferric Chloride 10%, Aniline 95%	N	Fluorocarbon	Carbon Graphite	Hast. "B"	Ceramic	Hastelloy

(Continued)

TABLE V (Cont'd)

Liquid	Lubricant?	Secondary Seals	Seal Washer	Metal Parts	Counterface	Spring(s)
Ferric Sulfate	N	Fluorocarbon Nitrile	Carbon Graphite	Hastelloy 316 SS	Ceramic	Hastelloy
Ferrous Sulfate & Lime	N	Fluorocarbon Nitrile	Carbon Graphite	316 SS	Ceramic	316 SS
Film Dope	N	Fluorocarbon	Carbon Graphite	316 SS	Ceramic	SS
Flit	N	Fluorocarbon	Carbon Graphite	316 SS	Stellite Stellite Face	SS
Fluorine	N	Fluorocarbon	Carbon Graphite	SS	Stellite	SS
Formaldehyde or Formalin	N	Fluorocarbon	Carbon Graphite	316 SS	Stellite Face Ceramic	316 SS
Freon 11	Y	Nitrile	Carbon Graphite	316 SS	Stellite Stellite Face Ni-Resist	316 SS
Freon 12	Y	Nitrile Chloroprene	Carbon Graphite	316 SS Steel	Cast Iron Stellite Face Ni-Resist	316 SS Steel
Freon 13	Y	Nitrile	Carbon Graphite	316 SS	Stellite Face	316 SS
Freon 14	Y	Fluorocarbon Chloroprene	Carbon Graphite	316 SS Steel	Cast Iron Stellite Face Bronze	316 SS
Freon 21	Y	Fluorocarbon Chloroprene	Carbon Graphite	316 SS Steel	Cast Iron Stellite Face	SS

Freon 22	Y	Fluorocarbon Chloroprene	Carbon Graphite	316 SS Steel	Stellite Cast Iron Stellite Face Bronze	316 SS
Freon 31	Y	Chloroprene	Carbon Graphite	316 SS Steel	Cast Iron Stellite Face	SS
Freon 32	Y	Chloroprene	Carbon Graphite	316 SS Steel	Cast Iron Stellite Face	SS
Freon 112	Y	Chloroprene * Nitrile	Carbon Graphite	316 SS Steel	Cast Iron Stellite Face	SS
Freon 113	Y	Chloroprene Fluorocarbon	Carbon Graphite	316 SS Steel	Cast Iron Stellite Face Bronze	316 SS
Freon 114	Y	Fluorocarbon Chloroprene Nitrile	Carbon Graphite	316 SS Steel	Cast Iron Stellite Face Bronze	316 SS
Freon 115	Y	Chloroprene Nitrile Fluorocarbon	Carbon Graphite	316 SS Steel	Cast Iron Stellite Face Bronze	316 SS
Freon 121	Y	Fluorocarbon Chloroprene	Carbon Graphite	316 SS Steel	Cast Iron Stellite Face	SS
Fruit Juices	N	Fluorocarbon Nitrile	Carbon Graphite	316 SS	Ceramic Ceramic Face	SS
Furfural	N	Fluorocarbon	Carbon Graphite	316 SS	Stellite Ceramic Stellite Face	316 SS

* Commonly specified by some seal manufacturers; however, more suitable materials may be available.

(Continued)

TABLE V (Cont'd)

Liquid	Lubricant?	Secondary Seals	Seal Washer	Metal Parts	Counterface	Spring(s)
Gasoline, Automotive Grades, Bronze, Ethyl, Hi-Test, 100-Octane, 130-Octane, Refined	Y	Nitrile	Carbon Graphite	316 SS Steel	Stellite Face Ni-Resist Stellite	316 SS
Gasoline, with H_2S, with Mercaptan	Y	Fluorocarbon	Carbon Graphite	316 SS	Stellite Stellite Face	316 SS
Gasoline – Aviation	Y	Fluoroelastomer Nitrile	Carbon Graphite	316 SS Steel	Ni-Resist Stellite Face	SS
Gasoline – Sour	Y	Fluorocarbon Nitrile	Carbon Graphite	316 SS	Stellite Ceramic Ni-Resist Stellite Face	SS
Gasoline – Tanker Service	Y	Nitrile	Carbon Graphite	Monel Steel	Ni-Resist Stellite Face	SS
Gasoline (Aromatic)		Fluorocarbon Fluoroelastomer	Carbon Graphite	316 SS	Ceramic Stellite Face Ni-Resist	SS
Gelatin	Y	Fluorocarbon Nitrile	Carbon Graphite	316 SS	Ceramic Stellite Face	SS
Glauber's Salt	See Sodium Sulfate Decahydrate					
Glucose	N	Nitrile	Carbon Graphite	316 SS Brass	Ceramic Ceramic Face	SS
Glue	N	Nitrile	Carbon Graphite	316 SS Brass	Ceramic Ceramic Face	SS

Glue Sizing	N	Nitrile Fluorocarbon	Carbon Graphite	316 SS Brass	Ceramic	SS
Glycerine or Glycerol	Y	Nitrile	Carbon Graphite	316 SS Steel	Ni-Resist Stellite Face	SS
Glycol, Diethylene See Diethylene Glycol						
Glycol, Ethylene See Ethylene Glycol						
Glycol, Propylene See Propylene Glycol						
Grain Mash—Wet	N	Fluorocarbon Nitrile	Carbon Graphite	316 SS	Ceramic Ceramic Face	SS
Green Sulfate Liquor	N	Fluorocarbon Fluoroelastomer	Carbon Graphite Ni-Resist	316 SS	Stellite Face Ceramic	SS
Helium		Nitrile	Carbon Graphite	316 SS	316 SS	316 SS
Heptane	Y	Fluoroelastomer Nitrile	Carbon Graphite	316 SS Steel	Stellite Face Ni-Resist	316 SS
Hexachloroacetone	Y	Fluorocarbon	Carbon Graphite	316 SS	Stellite Face	316 SS
Hexane	Y	Nitrile	Carbon Graphite	316 SS Steel	Ni-Resist Stellite Face	316 SS
Hexone See Isobutyl Methyl Ketone						
Hydrocarbons—Light		Nitrile	Carbon Graphite	316 SS	Stellite Face Tungsten Carbide	316 SS
Hydrogen		Nitrile	Carbon Graphite	316 SS	316 SS	316 SS
Hydrogen Bromide See Acid, Hydrobromic						
Hydrogen Cyanide See Acid, Hydrocyanic						

(Continued)

TABLE V (Cont'd)

Liquid	Lubricant?	Secondary Seals	Seal Washer	Metal Parts	Counterface	Spring(s)
Hydrogen Peroxide <30% <150°F	N	Fluorocarbon Fluoroelastomer	Carbon Graphite Glass Filled-Fluorocarbon	316 SS	Ceramic	SS
Hydrogen Sulfide		Nitrile	Carbon Graphite	316 SS	Ni-Resist	316 SS
Hydrogen Sulfide & Organic Sulfur Compounds	N	Fluorocarbon	Carbon Graphite	316 SS Monel	Ceramic	Monel
Hydrogen Sulfide Sol.—Dry & Cold	N	Fluorocarbon	Carbon Graphite	316 SS	Ceramic	316 SS
Hydrogen Sulfide Sol.—Wet & Cold	N	Fluorocarbon	Carbon Graphite	316 SS	Ceramic	316 SS
Hydrogen Sulfide Sol.—Wet & Hot	N	Fluorocarbon EPR	Carbon Graphite	Carp. 20	Ceramic	Carp. 20
Hydrogen and Water	N	Nitrile	Carbon Graphite	Brass	Ceramic	SS
Hydroquinone	N	Fluorocarbon	Carbon Graphite	SS	Ceramic	SS
Ink	N	Fluorocarbon	Carbon Graphite	316 SS	Ceramic	SS
Iodoform	N	Fluorocarbon	Carbon Graphite	Carp. 20 Hast. "C"	Stellite Ceramic	Carp. 20
Irish Moss Slurry	N	Fluorocarbon	Carbon Graphite	316 SS	Ceramic Stellite Face	SS
Iron Sulfate See Ferric Sulfate						
Isobutane	Y	Nitrile Chloroprene	Carbon Graphite	316 SS Steel	Stellite Ni-Resist Tungsten Carbide	316 SS

Isobutanol	See Alcohol, Isobutyl					
Isobutylene	Y	Nitrile Chloroprene	Carbon Graphite	316 SS Steel	Stellite Ni-Resist Tungsten Carbide	316 SS
Isobutyl Methyl Ketone (Hexone)	Y	Fluorocarbon Fluoroelastomer	Carbon Graphite	316 SS	Ceramic Stellite Face Bronze	316 SS
Isobutyraldehyde	Y	Fluorocarbon	Carbon Graphite	316 SS	Stellite Stellite Face	SS
Isopentane	Y	Chloroprene Nitrile	Carbon Graphite	316 SS Steel	Stellite Stellite Face Ni-Resist	316 SS
Isopropanol	See Alcohol, Isopropyl					
Isopropyl Acetate	N	Fluorocarbon	Carbon Graphite	316 SS	Ceramic Stellite Face Bronze	316 SS
Isopropyl Amine	Y	Fluorocarbon	Carbon Graphite	316 SS	Ceramic Stellite Face	SS
Jet Fuel JP3	Y	Fluoroelastomer	Carbon Graphite	316 SS Brass	Cast Iron Ni-Resist	SS
Jet Fuel JP4 or JP5	Y	Fluoroelastomer	Carbon Graphite	316 SS Brass	Stellite Stellite Face Cast Iron Ni-Resist	316 SS
Kaolin Slip Suspension in Water	N	Nitrile	Carbon Graphite	316 SS Brass	Ceramic Stellite Face	SS
Kerosene (Kerosine)	Y	Nitrile	Carbon Graphite	316 SS Steel	Stellite Ni-Resist Stellite Face	316 SS

(Continued)

TABLE V (Cont'd)

Liquid	Lubricant?	Secondary Seals	Seal Washer	Metal Parts	Counterface	Spring(s)
Ketchup	Y	Nitrile	Carbon Graphite	316 SS	Stellite Stellite Face	SS
Ketone in Oil	Y	Nitrile Fluorocarbon	Carbon Graphite	316 SS Steel	Ni-Resist Stellite Face	SS
Krypton		Nitrile	Carbon Graphite	316 SS	Ni-Resist	316 SS
Lacquers	N	Fluorocarbon	Carbon Graphite	316 SS	Ceramic Stellite Face	SS
Lacquer – MEK Used as Solvent	N	Fluorocarbon	Carbon Graphite	316 SS	Ceramic Stellite Face	SS
Lacquer Solvents	N	Fluorocarbon	Carbon Graphite	316 SS	Ceramic Stellite Face	SS
Lard	Y	Fluorocarbon Fluoroelastomer	Carbon Graphite	316 SS	Ceramic Stellite Face	SS
Lime – Sulfur	Y	Fluorocarbon	Carbon Graphite	316 SS	Stellite Face	316 SS
Lime Water (Milk of Lime)	Y	Nitrile	Carbon Graphite	316 SS	Ceramic Face Stellite Face	316 SS
Lindol See Tricresyl Phosphate						
Liquified Petroleum Gases (LPG)	N	Nitrile	Carbon Graphite	316 SS Brass	Tungsten Carbide	SS
Liquor, Black	N	Fluorocarbon Fluoroelastomer	Carbon Graphite Ni-Resist	316 SS	Ceramic Stellite Face	SS
Liquor, Green	N	Fluorocarbon Fluoroelastomer	Carbon Graphite Ni-Resist	316 SS	Ceramic Stellite Face	SS

Liquor, Lime	N	Fluorocarbon Nitrile	Carbon Graphite	316 SS	Ceramic Ceramic Face	SS
Liquor, Steep	N	Fluorocarbon Nitrile	Carbon Graphite	316 SS	Ceramic Ceramic Face	SS
Liquor, Sulfate	N	Fluorocarbon Fluoroelastomer	Carbon Graphite Ni-Resist	316 SS	Ceramic Stellite Face	SS
Liquor, Sulfite	N	Fluorocarbon Fluoroelastomer	Carbon Graphite	316 SS	Ceramic	SS
Liquor, White	N	Fluorocarbon Fluoroelastomer	Carbon Graphite	316 SS	Ceramic Stellite Face	SS
Lithium Bromide Brine	N	Fluorocarbon Chloroprene	Carbon Graphite	Brass Hast. "B"	Ceramic	SS
Lithium Chloride	N	Fluorocarbon Nitrile	Carbon Graphite	Hast. "B"	Ceramic	Hastelloy
Lithium Hydroxide	N	Fluorocarbon	Carbon Graphite	SS Carp. 20	Stellite Stellite Face	SS
Magnesium Acetate	N	Fluorocarbon	Carbon Graphite	Hast. "B"	Ceramic	Hast. "B"
Magnesium Chloride	N	Fluorocarbon	Carbon Graphite	Hast. "B"	Ceramic	Hastelloy
Magnesium Hydroxide	N	Fluorocarbon Nitrile *	Carbon Graphite	316 SS Carp. 20	Ceramic Stellite Face	Carp. 20
Magnesium Sulfate	N	Fluorocarbon Chloroprene	Carbon Graphite	316 SS	Ceramic Stellite Face	SS
Maleic Anhydride	N	Fluorocarbon Fluoroelastomer	Carbon Graphite	316 SS	Ceramic Ceramic Face	316 SS
Maleic Hydrazide	N	Fluorocarbon	Carbon Graphite	316 SS	Ceramic	316 SS

* Commonly specified by some seal manufacturers; however, more suitable materials may be available.

(Continued)

TABLE V (Cont'd)

Liquid	Lubricant?	Secondary Seals	Seal Washer	Metal Parts	Counterface	Spring(s)
Malt Beverages	N	Fluorocarbon Nitrile	Carbon Graphite	316 SS	Ceramic Stellite Face	SS
Manganese Chloride	N	Fluorocarbon Nitrile	Carbon Graphite	Hast. "B"	Ceramic	Hastelloy
Mash		Chloroprene Fluorocarbon	Carbon Graphite	316 SS	Ceramic	SS
Mayonnaise	N	Fluorocarbon Nitrile	Carbon Graphite	316 SS	Stellite Stellite Face	SS
MEK See Methyl Ethyl Ketone						
Melamine Resins	N	Fluorocarbon Nitrile	Carbon Graphite	316 SS	Ceramic Ceramic Face	SS
Mercaptan	N	Fluorocarbon	Carbon Graphite	316 SS	Stellite Stellite Face	SS
Mercuric Chloride – 10% at 160°F	N	Fluorocarbon Fluoroelastomer	Carbon Graphite	Hast. "C"	Ceramic	Hastelloy
Mercury	N	Fluorocarbon Nitrile	Carbon Graphite	316 SS	Ceramic Ceramic Face	SS
Mercury Salts	N	Fluorocarbon	Carbon Graphite	SS	Ceramic	SS
Mesityloxide	Y	Fluorocarbon Nitrile *	Carbon Graphite	316 SS	Stellite Stellite Face Bronze	316 SS
Methane		Nitrile	Carbon Graphite	316 SS	Tungsten Carbide	316 SS

Methanol See Alcohol, Methyl						
Methyl Acrylate	Y	Fluorocarbon	Carbon Graphite	316 SS	Ceramic Stellite Face	SS
Methyl Benzoate		Fluorocarbon	Carbon Graphite	316 SS	Bronze	316 SS
Methyl Bromide	N	Fluorocarbon	Carbon Graphite	Hast. "B" Monel SS	Stellite Ceramic	SS Hastelloy
Methyl Butyrate		Fluorocarbon	Carbon Graphite	316 SS	Bronze	316 SS
Methyl Cellosolve	N	Fluorocarbon	Carbon Graphite	316 SS	Ceramic Bronze Stellite Face	316 SS
Methyl Chloride	N	Fluorocarbon Fluoroelastomer	Carbon Graphite	316 SS Monel	Ceramic Stellite Face	SS Hastelloy
Methyl Cyclopentane	Y	Nitrile †	Carbon Graphite	316 SS	Ni-Resist 316 SS	316 SS
Methylene Chloride	N	Fluorocarbon	Carbon Graphite	316 SS Monel	Ceramic Stellite Face	Hastelloy SS
Methylene Dichloride See Methylene Chloride						
Methyl Ethyl Ketone (MEK)	N	Fluorocarbon EPR	Carbon Graphite	316 SS	Ceramic Stellite Face Bronze	316 SS
Methyl Formate		Fluorocarbon	Carbon Graphite	316 SS	Bronze	316 SS
Methyl Isobutyl Ketone	Y	Fluorocarbon	Carbon Graphite	316 SS	Ceramic Bronze Stellite Face	316 SS

* Commonly specified by some seal manufacturers; however, more suitable materials may be available.
† Limited data available.

(Continued)

TABLE V (Cont'd)

Liquid	Lubricant?	Secondary Seals	Seal Washer	Metal Parts	Counterface	Spring(s)
Methyl Methacrylate	N	Fluorocarbon	Carbon Graphite	316 SS	Ceramic Stellite Face Bronze	316 SS
Methyl Propionate		Fluorocarbon	Carbon Graphite	316 SS	Bronze	316 SS
Mineral Spirits (Painters' Naphtha)	Y	Fluorocarbon Nitrile	Carbon Graphite	316 SS Steel	Ni-Resist Stellite Face	316 SS
Molasses	Y	Nitrile Fluorocarbon Chloroprene	Carbon Graphite	316 SS Steel	Ni-Resist Stellite Face	SS
Mustard	N	Fluorocarbon Chloroprene	Carbon Graphite	316 SS Monel	Ceramic Stellite Face	Monel
Naphtha		Fluorocarbon Fluoroelastomer	Carbon Graphite	316 SS	Ceramic Stellite Face	316 SS
Naphtha Petroleum	Y	Nitrile	Carbon Graphite	316 SS Steel	Ni-Resist Stellite Face	SS
Naphtha – Sour	See Gasoline – Sour					
Naphthalene	N	Fluorocarbon Fluoroelastomer	Carbon Graphite	316 SS	Stellite Stellite Face	SS
Naphthalene – Molten		Fluorocarbon	Carbon Graphite	316 SS	Bronze	316 SS
Naphtha, UPM	Y	Fluorocarbon Nitrile	Carbon Graphite	316 SS Steel	Ni-Resist Stellite Face	SS
Neon		Nitrile	Carbon Graphite	316 SS	316 SS	316 SS

Nickel Chloride	N	Fluorocarbon	Carbon Graphite	Hast. "B"	Ceramic	Hast. "B"
Nickel Sulfate	N	Fluorocarbon Nitrile	Carbon Graphite	316 SS	Ceramic	SS
Nicotine Sulfate	N	Fluorocarbon	Carbon Graphite	316 SS	Stellite Ceramic	SS
Nitro Benzene	Y	Fluorocarbon	Carbon Graphite	316 SS	Stellite Ceramic	SS
Nitrogen Gas		Fluorocarbon	Carbon Graphite	316 SS	316 SS	316 SS
Nitro Methane	Y	Fluorocarbon Nitrile* Fluoroelastomer*	Carbon Graphite	316 SS	Ceramic Stellite Face 316 SS	316 SS
Oakite–Acetic	N	Fluorocarbon	Carbon Graphite	316 SS	Ceramic Ceramic Face	SS
Oakite–Alkaline	N	Fluorocarbon	Carbon Graphite	316 SS	Ceramic Ceramic Face	SS
Oakite Solvent	N	Fluorocarbon	Carbon Graphite	316 SS	Ceramic Ceramic Face	SS
Oil, Absorption	Y	Nitrile Fluoroelastomer	Carbon Graphite	316 SS Steel	Ni-Resist Stellite Face	SS
Oil, Angamol	Y	Nitrile Fluorocarbon	Carbon Graphite	316 SS Steel	Ni-Resist Stellite Face	SS
Oil, Air Corps Hydraulic	Y	Nitrile Fluoroelastomer	Carbon Graphite	316 SS Steel	Ni-Resist Stellite Face	SS
Oil, Apiezon	Y	Nitrile	Carbon Graphite	Steel	N-Resist	SS

* Commonly specified by some seal manufacturers; however, more suitable materials may be available.

(Continued)

TABLE V (Cont'd)

Liquid	Lubricant?	Secondary Seals	Seal Washer	Metal Parts	Counterface	Spring(s)
Oil, Bunker "C" Fuel	Y	Nitrile	Carbon Graphite Bronze	316 SS	Cast Iron Stellite Face	SS
Oil, Castor	Y	Fluorocarbon Nitrile	Carbon Graphite	316 SS	Ceramic Stellite Face	SS
Oil, Chinawood (Tung Oil)	Y	Fluorocarbon Nitrile	Carbon Graphite Bronze	316 SS	Ni-Resist Stellite Face	SS
Oil, Cocoanut	Y	Fluorocarbon Nitrile	Carbon Graphite	316 SS Steel	Ni-Resist Stellite Face	SS
Oil, Corn	Y	Nitrile	Carbon Graphite	316 SS Steel	Ni-Resist Stellite Face Stellite	316 SS
Oil, Cottonseed	Y	Nitrile	Carbon Graphite	316 SS Steel	Stellite Face Ni-Resist	SS
Oil, Crude–Asphalt Base (Petroleum Asphalt)	Y	Nitrile	Carbon Graphite Bronze	316 SS Steel	Ni-Resist Stellite Face	SS
Oil, Crude–Corrosive	Y	Nitrile Fluorocarbon	Carbon Graphite	316 SS	Stellite Ceramic Stellite Face	SS
Oil, Crude–Paraffin Base	Y	Fluorocarbon Chloroprene	Carbon Graphite	316 SS	Ceramic Stellite Stellite Face	SS
Oil, Crude–Sweet	Y	Nitrile	Carbon Graphite Bronze	316 SS	Ni-Resist Stellite Face	SS

Oil, Cutting, Soluble	Y	Nitrile	Carbon Graphite	316 SS Steel	Ni-Resist Stellite Face	SS
Oil, Diesel	Y	Nitrile Fluoroelastomer	Carbon Graphite	316 SS Steel	Ni-Resist 316 SS Stellite Face	316 SS
Oil, Domestic Fuel	Y	Nitrile	Carbon Graphite	316 SS Steel	Ni-Resist Stellite Face	SS
Oil, DTE	Y	Nitrile Fluorocarbon	Carbon Graphite	316 SS Steel	Ni-Resist Stellite Face	SS
Oil, Fuel & Gas	Y	Nitrile	Carbon Graphite	316 SS	Ni-Resist Stellite Face Stellite	316 SS
Oil, No. 6 Fuel		Nitrile	Carbon Graphite	316 SS	Cast Iron Stellite Face	SS
Oil, Fusel	Y	Fluorocarbon	Carbon Graphite	316 SS	Ceramic Stellite Face	SS
Oil, Hydraulic	Y	Nitrile Fluorocarbon	Carbon Graphite	316 SS Steel	Ni-Resist Stellite Face	SS
Oil, Hydraulic—Mineral Base	Y	Nitrile	Carbon Graphite	316 SS Steel	Ni-Resist Stellite Face	SS
Oil, Jet Fuel	Y	Nitrile Fluoroelastomer	Carbon Graphite	316 SS Steel	Ni-Resist	SS
Oil, Kerosene	See Kerosene					
Oil, Lean	Y	Nitrile Fluoroelastomer	Carbon Graphite	316 SS Steel	Ni-Resist Stellite Face	SS
Oil, Linseed	Y	Nitrile Fluorocarbon	Carbon Graphite	316 SS Steel	Ni-Resist Stellite Face	SS

(Continued)

TABLE V (Cont'd)

Liquid	Lubri-cant?	Secondary Seals	Seal Washer	Metal Parts	Counterface	Spring(s)
Oil, Lubricating–Refined		Nitrile	Carbon Graphite Bronze	316 SS	Stellite Stellite Face	316 SS
Oil, Mineral–Baby & Seal	Y	Nitrile	Carbon Graphite	316 SS Steel	Ni-Resist Stellite Face	SS
Oil, Mineral–Seal	Y	Nitrile	Carbon Graphite	316 SS Steel	Stellite Face Ni-Resist	SS
Oil, Nut	Y	Nitrile Fluorocarbon	Carbon Graphite	316 SS Steel	Stellite Face Ni-Resist	SS
Oil, Palm (Palm Butter)	Y	Nitrile Fluorocarbon	Carbon Graphite	316 SS Steel	Ni-Resist Stellite Face	SS
Oil, Paraffin Base	Y	Nitrile Fluorocarbon	Carbon Graphite	316 SS Steel	Ni-Resist Stellite Face	SS
Oil, Peanut	Y	Nitrile Fluorocarbon	Carbon Graphite	316 SS Steel	Ni-Resist Stellite Face	SS
Oil, Pine	Y	Fluorocarbon	Carbon Graphite	316 SS	Stellite Stellite Face	SS
Oil, Quenching	Y	Nitrile Fluorocarbon	Carbon Graphite	316 SS Steel	Ni-Resist Stellite Face	SS
Oil, Rapeseed	Y	Nitrile * Fluorocarbon	Carbon Graphite	316 SS Steel	Ni-Resist Stellite Face	SS
Oil, Rich	Y	Nitrile	Carbon Graphite	316 SS Steel	Ni-Resist Stellite Face	SS

Oil, Silicone	Y	Fluorocarbon	Carbon Graphite	316 SS	Ceramic Stellite Face	SS
Oil, Soya Bean	Y	Nitrile Chloroprene	Carbon Graphite	316 SS Steel	Ni-Resist Stellite Face	SS
Oil, Tall	Y	Nitrile Fluorocarbon	Carbon Graphite Bronze	316 SS Steel	Ni-Resist Stellite Face Stellite	316 SS
Oil, Transformer	Y	Fluoroelastomer Nitrile	Carbon Graphite	316 SS Steel	Ni-Resist Stellite Face	SS
Oil, Tung See Oil, Chinawood						
Oil, Ucon	Y	Fluorocarbon Nitrile	Carbon Graphite	316 SS	Cast Iron Ni-Resist	SS
Oil, Vegetable	Y	Nitrile Fluoroelastomer	Carbon Graphite	316 SS Steel	Ni-Resist Stellite Face	SS
Olefin – Crude	Y	Fluorocarbon Chloroprene Fluoroelastomer	Carbon Graphite	316 SS	Stellite Ceramic Stellite Face	SS
Oleum See Acid, Pyrosulfuric						
Oleum Spirits	Y	Fluorocarbon	Glass Filled- Fluorocarbon	Hastelloy Carp. 20	Ceramic Glass Filled- Fluorocarbon	Hastelloy
Ortho-dichlorobenzene	Y	Fluorocarbon Glass Filled- Fluorocarbon	Carbon Graphite	316 SS	Ceramic Stellite Face	SS
Ortho-dichlorobenzene – Pure		Fluorocarbon	Carbon Graphite	316 SS	Ni-Resist	316 SS
OS-45-1	Y	Fluorocarbon	Carbon Graphite	SS	Ceramic	SS

[*] Commonly specified by some seal manufacturers; however, more suitable materials may be available.

(Continued)

TABLE V (Cont'd)

Liquid	Lubricant?	Secondary Seals	Seal Washer	Metal Parts	Counterface	Spring(s)
Oxygen		Nitrile *	Carbon Graphite	316 SS	316 SS	316 SS
Paint	N	Fluorocarbon	Carbon Graphite	316 SS	Ceramic Ni-Resist	SS
Para-cymene	Y	Fluorocarbon Fluoroelastomer	Carbon Graphite	316 SS	Stellite Stellite Face Ceramic	316 SS
Paraffin – Molten, <170°F	Y	Nitrile Glass Filled-Fluorocarbon	Carbon Graphite	316 SS Steel	Ni-Resist	SS
Paraffin – Molten, >170°F	Y	Fluorocarbon	Carbon Graphite	SS	Stellite	SS
Paraffin Wax 60%		Nitrile Glass Filled-Fluorocarbon	Carbon Graphite	316 SS Steel	Ni-Resist	SS
Pectin Liquor	N	Fluorocarbon Chloroprene Nitrile	Carbon Graphite	316 SS	Ceramic Stellite Face	SS
Penicillin – Liquid	N	Fluoroelastomer Nitrile	Carbon Graphite	316 SS	Stellite Ceramic	SS
Pentachlorophenol		Fluorocarbon	Carbon Graphite	SS Hast. "B"	Ceramic	SS
Pentane Liquid	N	Nitrile Chloroprene	Carbon Graphite	316 SS Steel	Tungsten Carbide Stellite Stellite Face	316 SS

Pentasol	Y	Fluorocarbon	Glass Filled-Fluorocarbon	Hastelloy	Ceramic	Hastelloy
Perchloroethylene – Clean		Fluorocarbon	Carbon Graphite	316 SS	Bronze	316 SS
Perchloroethylene – Contaminated		Fluorocarbon	Fluorocarbon Asbestos	316 SS	Ceramic	SS-Fluorocarbon Lined
Perchloroethylene – Dry-Cleaning		Fluorocarbon	Carbon Graphite	316 SS	Ceramic	316 SS
Perchloroethylene – Max. Temp. 170°F	Y	Fluorocarbon	Carbon Graphite	SS Carp. 20	Ceramic	SS
Perchloroethylene & Filter Acid	N	Fluorocarbon	Carbon Graphite	SS Carp. 20	Ceramic	SS
Petroleum Ether	See Benzine (Ligroin)					
Petroleum Spirit – <20% Aromatics		Fluorocarbon	Carbon Graphite	316 SS	316 SS	316 SS
Petroleum Spirit – >20% Aromatics		Fluorocarbon	Carbon Graphite	316 SS	Bronze	316 SS
Phenol	N	Fluorocarbon	Carbon Graphite	316 SS	Stellite Stellite Face Ceramic	316 SS
Phenol – Formaldehyde Mix	N	Fluorocarbon	Carbon Graphite	316 SS	Stellite Stellite Face	SS
Phenol – 10% & Water	N	Fluorocarbon Fluoroelastomer	Carbon Graphite	316 SS	Stellite Stellite Face Ceramic	316 SS
Phosphorous Trichloride	N	Fluorocarbon	Carbon Graphite	Hast. "C"	Ceramic Ceramic Face	Hast. "C"

* Commonly specified by some seal manufacturers; however, more suitable materials may be available.

(Continued)

TABLE V (Cont'd)

Liquid	Lubricant?	Secondary Seals	Seal Washer	Metal Parts	Counterface	Spring(s)
Photographic Developers	N	Fluorocarbon	Carbon Graphite	316 SS	Ceramic Stellite Face	SS
Phthalate Esters	N	Fluorocarbon Glass Filled-Fluorocarbon	Carbon Graphite	316 SS	Stellite Stellite Face	SS
Plasticizer (Liquid Dibutyl Cellosolve Adipate)	N	Fluorocarbon	Carbon Graphite	316 SS	Ceramic Stellite Face	SS
Plating Sol.—Antimony		Chloroprene	Carbon Graphite	316 SS	Ceramic	316 SS
Plating Sol.—Arsenic		Chloroprene	Carbon Graphite	316 SS	Ceramic	316 SS
Plating Sol.—Brass Regular Brass Bath		Chloroprene	Carbon Graphite	Hast. "B"	Ceramic	Hast. "B"
High Speed Brass Bath		Nitrile	Carbon Graphite	316 SS	Ceramic	316 SS
Plating Sol.—Bronze Copper-Cadmium Bronze Bath		Chloroprene	Carbon Graphite	316 SS Hast. "B"	Ceramic	316 SS Hast. "B"
Copper-Tin Bronze Bath		Nitrile	Carbon Graphite	316 SS	Ceramic	316 SS
Copper-Zinc Bronze Bath		Chloroprene	Carbon Graphite	316 SS	Ceramic	316 SS
Plating Sol.—Cadmium Cyanide Bath		Chloroprene	Carbon Graphite	Hast. "B"	Ceramic Carp. 20	Hast. "B"
Fluoborate Bath		Chloroprene	Carbon Graphite	Hast. "B"	Carp. 20	Hast. "B"

Plating Sol.—Chromium Chromic-Sulfuric Bath	Fluoroelastomer	Carbon Graphite	N/A	Ceramic	N/A
Fluosilicate Bath	Fluoroelastomer	Carbon Graphite	Carp. 20	Ceramic Carp. 20	Carp. 20
Fluoride Bath	Fluoroelastomer	Carbon Graphite	Carp. 20	Ceramic Carp. 20	Carp. 20
Black Chrome Bath	Nitrile Fluoroelastomer	Carbon Graphite	N/A	Ceramic	N/A
Barrel Chrome Bath	Fluoroelastomer	Carbon Graphite	N/A	Ceramic	N/A
Plating Sol.—Copper (Acid) Copper Sulfate Baths	Nitrile Fluoroelastomer	Carbon Graphite	Hast. "B"	Ceramic	Hast. "B"
Copper Fluoborate Baths	Chloroprene	Carbon Graphite	Hast. "B"	Carp. 20	Hast. "B"
Plating Sol.—Copper (Cyanide) Copper Strike Bath	Chloroprene	Carbon Graphite	316 SS	Ceramic	316 SS
Rochelle & Potassium Salt Baths	Chloroprene	Carbon Graphite	316 SS	Ceramic	316 SS
Rochelle & Sodium Salt Baths	Chloroprene	Carbon Graphite	Hast. "B"	Ceramic	Hast. "B"
Barrel Copper Bath	Nitrile	Carbon Graphite	316 SS	Ceramic	316 SS
Plating Sol.—Gold	Chloroprene	Carbon Graphite	316 SS	Ceramic	316 SS
Plating Sol.—Indium Cyanide	Nitrile	Carbon Graphite	316 SS	Ceramic	316 SS

N/A—not available.

(Continued)

TABLE V (Cont'd)

Liquid	Lubricant?	Secondary Seals	Seal Washer	Metal Parts	Counterface	Spring(s)
Plating Sol.—Iron Chloride Bath		Chloroprene	Carbon Graphite	N/A	Ceramic	N/A
Ferrous Sulfate Bath		Chloroprene	Carbon Graphite	316 SS	Ceramic	316 SS
Ferrous Ammonium Sulfate Bath		Chloroprene	Carbon Graphite	Hast. "B"	Ceramic	Hast. "B"
Sulfate-Chloride Bath		Chloroprene	Carbon Graphite	N/A	Ceramic	N/A
Fluoborate Bath		Chloroprene	Carbon Graphite	Hast. "B"	Carp. 20	Hast. "B"
Plating Sol.—Lead Fluoborate		Chloroprene	Carbon Graphite	Hast. "B"	Carp. 20	Hast. "B"
Plating Sol.—Nickel Nickel Fluoborate Bath		Chloroprene	Carbon Graphite	Hast. "B"	Carp. 20	Hast. "B"
All Other Nickel Baths		Nitrile	Carbon Graphite	N/A	Ceramic	N/A
Plating Sol.—Silver		Chloroprene	Carbon Graphite	316 SS	Ceramic	316 SS
Plating Sol.—Tin-Fluoborate		Chloroprene	Carbon Graphite	Hast. "B"	Carp. 20	Hast. "B"
Plating Sol.—Tin-Nickel		Nitrile	Carbon Graphite	Hast. "B"	Carp. 20	Hast. "B"
Plating Sol.—Zinc Acid Sulfate Bath		Chloroprene Nitrile	Carbon Graphite	Hast. "B"	Ceramic	Hast. "B"
Acid Fluoborate Bath		Chloroprene	Carbon Graphite	Hast. "B"	Carp. 20	Hast. "B"

Alkaline Zincate Bath		Nitrile	Carbon Graphite	Hast. "B"	Carp. 20	Hast. "B"
Cyanide Bath		Chloroprene	Carbon Graphite	Hast. "B"	Carp. 20	Hast. "B"
Poly Glycols	Y	Nitrile	Carbon Graphite	316 SS Brass	Ceramic Stellite Face	SS
Polyvinyl Acetates	N	Fluorocarbon	Carbon Graphite	Carp. 20 316 SS	Ceramic Stellite Face	Carp. 20
Polyvinyl Chloride	N	Fluorocarbon	Carbon Graphite	Carp. 20	Ceramic	Carp. 20
Potash Alum (Aluminum Potassium Sulfate)	N	Fluorocarbon	Carbon Graphite Tungsten Carbide	Carp. 20 Monel	Stellite Solid Tungsten Carbide	SS
Potassium Bicarbonate	N	Fluorocarbon Nitrile	Carbon Graphite	316 SS	Stellite Ceramic	SS
Potassium Bromide	N	Fluorocarbon Nitrile	Carbon Graphite	Hastelloy Monel	Ceramic	Hastelloy
Potassium Carbonate	N	Fluorocarbon Nitrile	Carbon Graphite	316 SS	Ceramic Tungsten Carbide	SS
Potassium Chlorate	N	Fluorocarbon Nitrile	Carbon Graphite	Hastelloy Monel	Ceramic Ceramic Face	Hastelloy
Potassium Chloride		Nitrile	Carbon Graphite	Monel	Ceramic Ceramic Face	Monel
Potassium Cyanide	N	Fluorocarbon Nitrile	Carbon Graphite	316 SS	Ceramic Ceramic Face	SS
Potassium Hydroxide	N	Fluorocarbon	Tungsten Carbide	Monel	Solid Tungsten Carbide	Monel

N/A – not available.

(Continued)

TABLE V (Cont'd)

Liquid	Lubricant?	Secondary Seals	Seal Washer	Metal Parts	Counterface	Spring(s)
Potassium Hydroxide – 50% Aqueous Sol.	N	Fluorocarbon	Tungsten Carbide	Monel	Tungsten Carbide	Monel
Potassium Hydroxide – Above 50%	N	Fluorocarbon	Tungsten Carbide	Monel	Tungsten Carbide	Monel
Potassium Nitrate	N	Fluorocarbon Nitrile	Carbon Graphite	316 SS	Ceramic	SS
Potassium Perfluoroacetate	N	Fluorocarbon	Carbon Graphite	SS Carp. 20	Ceramic	SS
Potassium Permanganate	N	Fluorocarbon	Carbon Graphite	316 SS	Ceramic Ceramic Face	SS
Potassium Phosphate – Di or Tri	N	Fluorocarbon Nitrile	Carbon Graphite	316 SS	Stellite Ceramic	SS
Potassium Silicate	N	Fluorocarbon	Carbon Graphite	SS Monel	Stellite Stellite Face	SS
Potassium Sulfate	N	Nitrile Fluorocarbon	Carbon Graphite	SS Monel	Ceramic Ceramic Face	SS
Preston Producer Gas	N	Fluorocarbon	Carbon Graphite	316 SS	Stellite Stellite Face	SS
Propane	Y	Fluorocarbon Nitrile	Carbon Graphite	316 SS	Stellite Stellite Face Ni-Resist Tungsten Carbide	316 SS
Propanol	See Alcohol, Propyl					

Propionaldehyde	Y	Fluorocarbon	Carbon Graphite	316 SS	Stellite Ceramic Ceramic Face	316 SS
Propyl Acetate		Fluorocarbon	Carbon Graphite	316 SS	316 SS	316 SS
Propyl Benzoate		Fluorocarbon	Carbon Graphite	316 SS	316 SS	316 SS
Propyl Butyrate		Fluorocarbon	Carbon Graphite	316 SS	316 SS	316 SS
Propylene		Nitrile *	Carbon Graphite	316 SS	Tungsten Carbide	316 SS
Propylene Glycol	Y	Nitrile	Carbon Graphite	316 SS Steel	Ni-Resist Stellite Face	SS
Propylene Oxide	Y	Fluorocarbon	Carbon Graphite	316 SS	Stellite Stellite Face	SS
Propylene Polymer	Y	Fluorocarbon Nitrile	Carbon Graphite	316 SS Steel	Ni-Resist Stellite Face	SS
Propyl Formate		Fluorocarbon	Carbon Graphite	316 SS	316 SS	316 SS
Propyl Propionate		Fluorocarbon	Carbon Graphite	316 SS	316 SS	316 SS
Protein Slurry	N	Fluorocarbon	Carbon Graphite	316 SS	Ceramic	SS
Pulp Stock (Paper)	N	Fluorocarbon	Carbon Graphite	316 SS	Ceramic	SS
Pyridine	N	Fluorocarbon	Carbon Graphite	316 SS	Stellite Stellite Face 316 SS	316 SS
Raffinate	Y	Fluorocarbon	Carbon Graphite	316 SS	Stellite Stellite Face	SS
Resorcinol Paper Mill	N	Fluorocarbon	Carbon Graphite	316 SS Carp. 20	Stellite Stellite Face	Carp. 20

* Commonly specified by some seal manufacturers; however, more suitable materials may be available.

(Continued)

TABLE V (Cont'd)

Liquid	Lubricant?	Secondary Seals	Seal Washer	Metal Parts	Counterface	Spring(s)
Rosins	N	Fluorocarbon	Carbon Graphite	SS	Ceramic	SS
Sal Ammoniac See Ammonium Chloride						
Sal Soda See Sodium Carbonate						
Salt Cake (Impure Sodium Sulfate)		Chloroprene Fluorocarbon Nitrile	Carbon Graphite	316 SS	Ni-Resist Ceramic	SS
Sewage	N	Nitrile	Carbon Graphite Bronze	316 SS	Ceramic Stellite Ceramic Face Stellite Face	SS
Sewage Sludge	N	Nitrile	Carbon Graphite	316 SS Brass	Ceramic Ceramic Face	SS
Silicone Fluid	Y	Fluorocarbon	Carbon Graphite	316 SS	Ceramic Stellite Face	SS
Silicone Tetrachloride	Y	Fluorocarbon	Carbon Graphite	316 SS	Stellite Face Stellite Ceramic	316 SS
Silver Nitrate – 10%	N	Fluorocarbon	Carbon Graphite	316 SS	Stellite Ceramic	SS
Slop, Brewery	N	Fluorocarbon	Carbon Graphite	316 SS	Ceramic Ceramic Face	SS
Slop, Distillers	N	Fluorocarbon	Carbon Graphite	316 SS	Ceramic Ceramic Face	SS

Soap, Liquor	Y	Fluorocarbon Nitrile	Carbon Graphite	316 SS Monel	Ceramic Tungsten Carbide	Monel
Soap Solutions		Fluorocarbon	Carbon Graphite	316 SS	Ceramic Face	316 SS
Soda Ash See Sodium Carbonate						
Soda, Caustic (Sodium Hydroxide)		Nitrile	Carbon Graphite	316 SS	Ceramic	316 SS
Sodium Acetate	N	Fluorocarbon	Carbon Graphite	316 SS	Ceramic Stellite Face	SS
Sodium Aluminate	N	Fluorocarbon	Carbon Graphite	Monel	Ceramic Ceramic Face	Monel
Sodium Aluminate Lime	N	Fluorocarbon	Carbon Graphite	Monel	Ceramic Ceramic Face	Monel
Sodium Bicarbonate (Baking Soda)	N	Fluorocarbon Chloroprene Nitrile	Carbon Graphite	316 SS	Stellite Ceramic Face Ceramic	SS
Sodium Bisulfate	N	Fluorocarbon Nitrile	Carbon Graphite	Hastelloy Monel	Ceramic Ceramic Face	Hastelloy
Sodium Bisulfite	N	Fluorocarbon Chloroprene Nitrile	Carbon Graphite	SS Monel	Stellite Ceramic Face	SS
Sodium Carbonate (Washing Soda, Sal Soda, Soda Ash)	N	Fluorocarbon Nitrile	Carbon Graphite	316 SS Brass	Ceramic Stellite Face	SS
Sodium Chloride	N	Chloroprene Nitrile	Carbon Graphite	Monel 316 SS	Ceramic	Monel
Sodium Chloride Brine	N	Fluorocarbon Nitrile	Carbon Graphite	316 SS Brass	Ceramic	316 SS

(Continued)

TABLE V (Cont'd)

Liquid	Lubricant?	Secondary Seals	Seal Washer	Metal Parts	Counterface	Spring(s)
Sodium Cyanamide	N	Fluorocarbon	Carbon Graphite	316 SS	Ceramic Stellite Face	SS
Sodium Cyanide	N	Fluorocarbon Nitrile	Carbon Graphite	316 SS	Ceramic Ceramic Face	SS
Sodium Dichromate & Water	N	Fluorocarbon	Carbon Graphite	316 SS	Ceramic	316 SS
Sodium Hydrosulfite	N	Fluorocarbon	Carbon Graphite	316 SS	Stellite Ceramic	SS
Sodium Hydroxide (Caustic Soda – Lye)		Chloroprene	Carbon Graphite Tungsten Carbide	Monel 316 SS	Tungsten Carbide	Monel 316 SS
Sodium Hypochlorite – Up to 20%, Up to 100°F	N	Fluorocarbon	Carbon Graphite	Hast. "C"	Ceramic	Hastelloy
Sodium Metaphosphate	N	Fluorocarbon	Carbon Graphite	316 SS	Ceramic	SS
Sodium Nitrate	N	Fluorocarbon Nitrile *	Carbon Graphite Glass Filled-Fluorocarbon	316 SS	Ceramic Ceramic Face	SS
Sodium Perborate	N	Fluorocarbon Nitrile *	Carbon Graphite	316 SS	Ceramic Ceramic Face	SS
Sodium Peroxide Sol.	N	Fluorocarbon	Carbon Graphite	316 SS	Ceramic Glass Filled-Fluorocarbon	SS
Sodium Phosphate, Mono-		Nitrile	Carbon Graphite	316 SS	Stellite Face Ceramic	316 SS

Sodium Phosphate, Di-	N	Fluorocarbon Nitrile	Carbon Graphite	316 SS	Ceramic Stellite Face	SS
Sodium Phosphate, Tri-	N	Nitrile	Carbon Graphite	316 SS Steel	Ni-Resist Ceramic Stellite Face	SS
Sodium Plumbite (Doctor's Sol.)	N	Fluorocarbon	Carbon Graphite	316 SS	Ceramic Stellite Face	SS
Sodium Silicate	N	Nitrile	Carbon Graphite	316 SS Brass	Ceramic Stellite Face	SS
Sodium Sulfate Decahydrate (Glauber's Salts)	N	Chloroprene Fluorocarbon Nitrile	Carbon Graphite	SS Monel	Ni-Resist Ceramic Ceramic Face	SS
Sodium Sulfate – Molten	N	Fluorocarbon Chloroprene Nitrile	Carbon Graphite	SS Monel	Ceramic Ceramic Face	SS
Sodium Sulfide	N	Fluorocarbon Nitrile	Carbon Graphite	316 SS	Stellite Ceramic Ceramic Face	316 SS
Sodium Sulfite	N	Fluorocarbon Nitrile	Carbon Graphite	316 SS	Stellite Ceramic Face	SS
Sodium Tetraborate (Borax)	N	Fluorocarbon Chloroprene	Carbon Graphite	316 SS	Stellite Stellite Face	SS
Sodium Thiosulfate ("Hypo")	N	Fluorocarbon Nitrile *	Carbon Graphite	316 SS	Ceramic Ceramic Face	SS
Solvasol – <150°F	N	Fluorocarbon Nitrile	Carbon Graphite	316 SS Steel	Ni-Resist Stellite Face	SS
Solvasol – >150°F	N	Fluorocarbon	Carbon Graphite	316 SS	Ceramic Stellite Face	SS

* Commonly specified by some seal manufacturers; however, more suitable materials may be available.

(Continued)

TABLE V (Cont'd)

Liquid	Lubricant?	Secondary Seals	Seal Washer	Metal Parts	Counterface	Spring(s)
Stannic Chloride–10%, 160°F	N	Nitrile	Carbon Graphite	Hast. "B"	Ceramic Stellite Face	Hast. "B"
Starch Slurry	N	Nitrile	Carbon Graphite	316 SS Brass	Ceramic Ceramic Face	SS
Starch, Thixotropic Sol.	N	Nitrile	Carbon Graphite	316 SS Steel	Ni-Resist Stellite Face	SS
Stoddard Solvent	Y	Nitrile Fluorocarbon	Carbon Graphite	316 SS Steel	Ni-Resist Ceramic Face	SS
Stoddard Solvent–W/Filter Aid	N	Nitrile Fluorocarbon	Carbon Graphite	316 SS Steel	Ni-Resist Ceramic Face	SS
Styrene	N	Fluorocarbon	Carbon Graphite	316 SS	316 SS Ceramic Stellite Face	316 SS
Styrene & Dinitro Chloro-benzene	N	Fluorocarbon	Carbon Graphite	316 SS	Ceramic Ni-Resist	SS
Sugar Solutions	N	Fluorocarbon Chloroprene Nitrile	Carbon Graphite	316 SS	Stellite Ceramic Ceramic Face	SS
Sulfahydrate	N	Fluorocarbon	Carbon Graphite	Hast. "C"	Ceramic	Hastelloy
Sulfate, Liquors–Kraft		Fluorocarbon	Ni-Resist	316 SS	Stellite Face	316 SS
Sulfite Pulp (Wood Pulp)	N	Fluorocarbon	Carbon Graphite	Hast. "C"	Ceramic	Hast. "C"
Sulfonated Vegetable Oils	Y	Fluorocarbon	Carbon Graphite	316 SS	Stellite Stellite Face	SS

Sulfur Chloride	N	Fluorocarbon Glass Filled-Fluorocarbon	Ni-Resist Fluorocarbon-Asbestos	316 SS	Stellite Face Ceramic	SS-Flucrocarbon Lined
Sulfur Dioxide		Fluorocarbon	Carbon Graphite	316 SS	316 SS	316 SS
Sulfur Dioxide – Wet		Fluorocarbon Fluoroelastomer	Carbon Graphite	Hast. "C"	Ceramic Ceramic Face	Hast. "C"
Sulfur Dioxide, Anhydrous Liquid – Cryogenic	N	Fluorocarbon	Carbon Graphite	Hast. "C" 316 SS	Ceramic	316 SS
Sulfur – Molten	N	Fluorocarbon Glass Filled-Fluorocarbon	Carbon Graphite Ni-Resist	316 SS	Ceramic Stellite Face	SS
Sulfur Trioxide	N	Fluorocarbon	Glass Filled-Fluorocarbon Carbon Graphite	Hast. "C"	Ceramic	Hastelloy
Syrup (High Sugar) <200°F	N	Fluorocarbon Nitrile	Carbon Graphite	316 SS	Ceramic	SS
Syrup (High Sugar) >200°F	N	Fluorocarbon	Carbon Graphite	316 SS	Ceramic	SS
Tallow – Hot	Y	Nitrile Glass Filled-Fluorocarbon	Carbon Graphite	316 SS Steel	Ni-Resist	SS
Tanning Liquors – Veg.	N	Fluorocarbon	Carbon Graphite	316 SS	Stellite Stellite Face	SS
Tar & Ammonia	N	Fluorocarbon Glass Filled-Fluorocarbon	Carbon Graphite	316 SS	Ceramic Stellite Face	SS

(Continued)

TABLE V (Cont'd)

Liquid	Lubricant?	Secondary Seals	Seal Washer	Metal Parts	Counterface	Spring(s)
Tar, Hot	N	Fluorocarbon Glass Filled-Fluorocarbon	Carbon Graphite	316 SS	Ceramic Stellite Face	SS
Tetrachloroethylene (Perchloroethylene)	N	Fluorocarbon	Carbon Graphite	316 SS Monel	Stellite Ceramic Stellite Face	Monel Hastelloy
Tetrachloroethane	N	Fluorocarbon	Carbon Graphite	316 SS	Ceramic Stellite Face	SS
Tetraethyl Lead	N	Fluorocarbon Chloroprene *	Carbon Graphite	316 SS	Ceramic Stellite Stellite Face Ni-Resist	316 SS
Tetrahydrofuran	Y	Fluorocarbon	Carbon Graphite	316 SS	Ceramic Stellite Face 316 SS	316 SS
Titanium Tetrachloride	N	Fluorocarbon Fluoroelastomer	Carbon Graphite Fluorocarbon-Asbestos	316 SS Monel Carp. 20	Ceramic	Monel SS-Fluorocarbon Lined
Toluene or Toluol	Y	Fluorocarbon Fluoroelastomer	Carbon Graphite	316 SS	Ni-Resist Ceramic Stellite Face	316 SS
Tomato Juice or Pulp	N	Fluorocarbon Chloroprene Nitrile	Carbon Graphite	316 SS	Ceramic Stellite Face	SS
Tooth Paste	Y	Nitrile	Carbon Graphite	316 SS	Ceramic Stellite Face	SS

Toxathene	Y	Fluorocarbon	Carbon Graphite	316 SS Carp. 20	Ceramic Stellite Face	Carp. 20
Tricresyl Phosphate (Lindol)	Y	Fluorocarbon	Carbon Graphite	316 SS	Ceramic Stellite Face	SS
Triethylamine	Y	Fluorocarbon Fluoroelastomer Nitrile	Carbon Graphite	316 SS Steel	Stellite Face Ni-Resist 316 SS	316 SS
Triethanolamine	Y	Fluorocarbon	Carbon Graphite	316 SS	Stellite Face Ceramic	SS
Trifluorovinyl-Chloride (Chlorotrifluoroethylene)	N	Fluorocarbon	Carbon Graphite	Carp. 20 Hast. "C"	Ceramic	Carp. 20
Turpentine	Y	Fluorocarbon Fluoroelastomer	Carbon Graphite	316 SS	Stellite Stellite Face Bronze	316 SS
Ucon Hydrolubes	See Oil, Ucon					
Urea–Phenolic Resins	N	Fluorocarbon	Carbon Graphite	316 SS	Ceramic Stellite Face	SS
Varnish–Aromatic	N	Fluorocarbon	Carbon Graphite	316 SS	Ceramic Stellite Face	SS
Varnish–Non-aromatic	N	Fluoroelastomer Nitrile*	Carbon Graphite	316 SS Steel	Ni-Resist Stellite Face	SS
Vegetable Juices	N	Fluorocarbon	Carbon Graphite	316 SS	Stellite Stellite Face	SS
Vegetable Oils	Y	Fluorocarbon	Carbon Graphite	316 SS	Ceramic Ceramic Face	SS
Vinegar	N	Fluorocarbon Fluoroelastomer	Carbon Graphite	316 SS	Ceramic	SS

* Commonly specified by some seal manufacturers; however, more suitable materials may be available.

(Continued)

TABLE V (Cont'd)

Liquid	Lubricant?	Secondary Seals	Seal Washer	Metal Parts	Counterface	Spring(s)
Vinyl Acetate	N	Fluorocarbon	Carbon Graphite	316 SS	Stellite Stellite Face	SS
Vinyl Chloride – Dry	N	Fluorocarbon	Carbon Graphite	316 SS	Stellite Stellite Face	316 SS
Vinyl Chloride – Wet or Impure	N	Fluorocarbon	Carbon Graphite	Hast. "C"	Ceramic	Hastelloy
Vinyl Pyridine	N	Fluorocarbon	Carbon Graphite	316 SS	Stellite Stellite Face	SS
Vinylidene Chloride	N	Fluorocarbon	Carbon Graphite	316 SS	Ceramic	SS
Washing Soda	See Sodium Carbonate					
Water, Boiler Feed		Fluoroelastomer	Carbon Graphite	316 SS	Tungsten Carbide	316 SS
Water, Brackish		Fluoroelastomer	Carbon Graphite	Monel	Bronze	Hastelloy
Water, Chilled Air Washing		Fluoroelastomer	Carbon Graphite	316 SS	Bronze	316 SS
Water, Chilled Condenser Service		Fluoroelastomer	Carbon Graphite	316 SS	Bronze	316 SS
Water Condensate Under 210°F		Nitrile	Carbon Graphite	316 SS	Stellite	316 SS
Water, Cooling Tower		Nitrile	Carbon Graphite	316 SS	Ceramic Face	316 SS
Water, De-ionized	N	Nitrile Fluoroelastomer	Carbon Graphite	316 SS	Ceramic Face Ceramic Bronze	316 SS

Water, De-mineralized		Fluorocarbon	Carbon Graphite	316 SS	Bronze	316 SS
Water, Detergent		Fluoroelastomer	Carbon Graphite	316 SS	Bronze	316 SS
Water, Detergent– pH less than 10	N	Nitrile	Carbon Graphite	316 SS	Ceramic Ceramic Face	SS
Water, Distilled	N	Fluorocarbon Nitrile	Carbon Graphite	316 SS	Ceramic Ceramic Face Bronze	316 SS
Water, Distilled–Cool		Nitrile	Carbon Graphite	316 SS	Bronze	316 SS
Water, Drinking		Fluorocarbon	Carbon Graphite	316 SS	Bronze	316 SS
Water, Fresh (Clean, Cool)	N	Nitrile	Carbon Graphite	316 SS	Ceramic Ceramic Face	SS
Water, Grit Carrying, Cool	N	Nitrile Fluoroelastomer	Carbon Graphite Tungsten Carbide	316 SS	Ceramic Ceramic Face Tungsten Carbide	316 SS
Water, Mine		Fluorocarbon Fluoroelastomer	Carbon Graphite	316 SS	Ceramic Face Tungsten Carbide	316 SS
Water, Salt	N	Fluorocarbon Nitrile	Carbon Graphite	Monel 316 SS	Ceramic	Monel
Water, Sandy–1%		Fluoroelastomer	Carbon Graphite	Monel	Bronze	Hastelloy
Water, Sandy–10%		Fluoroelastomer	Tungsten Carbide	316 SS	Tungsten Carbide	316 SS
Water, Sea	N	Fluorocarbon Fluoroelastomer Nitrile	Carbon Graphite	Monel	Ceramic Ceramic Face Bronze	Monel Hastelloy
Water, Sea Brine	N	Fluorocarbon	Carbon Graphite	316 SS Monel	Ceramic Ceramic Face	Monel

(Continued)

TABLE V (Cont'd)

Liquid	Lubricant?	Secondary Seals	Seal Washer	Metal Parts	Counterface	Spring(s)
Water, Soapy	Y	Fluorocarbon Chloroprene Nitrile Fluoroelastomer	Carbon Graphite	316 SS	Ceramic Ceramic Face Bronze	316 SS
Water with Soluble Oil	Y	Nitrile	Carbon Graphite	316 SS	Ceramic Ceramic Face	SS
Wax Slurry	Y	Fluorocarbon	Carbon Graphite	316 SS	Ceramic Ceramic Face	SS
Whey—Condensed	Y	Fluorocarbon Nitrile	Carbon Graphite	316 SS	Ceramic	SS
Whisky	Y	Fluorocarbon Fluoroelastomer Nitrile	Carbon Graphite	316 SS	Stellite Stellite Face Ni-Resist	316 SS
White Spirit		Fluorocarbon	Carbon Graphite	316 SS	Bronze	316 SS
White Water	N	Fluorocarbon Fluoroelastomer	Carbon Graphite	316 SS	Ceramic Stellite Face	SS
Wine	N	Fluorocarbon Nitrile	Carbon Graphite	316 SS	Ceramic Stellite Face	SS
Xenon		Nitrile	Carbon Graphite	316 SS	316 SS	316 SS
Xylene (Xylol, Xylole)	Y	Fluorocarbon Fluoroelastomer	Carbon Graphite	316 SS	Stellite Stellite Face Ni-Resist	316 SS
Xylene, Meta		Fluorocarbon	Carbon Graphite	316 SS	Ni-Resist	316 SS

Xylene, Ortho		Fluorocarbon	Carbon Graphite	316 SS	Ni-Resist	316 SS
Xylene, Para		Fluorocarbon	Carbon Graphite	316 SS	Ni-Resist	316 SS
Zeolite Treated Water	Y	Fluorocarbon Nitrile	Carbon Graphite	316 SS	Ceramic Ceramic Face	SS
Zinc Ammonium Chloride	N	Fluorocarbon Nitrile	Carbon Graphite	Carp. 20 Hast. "B"	Ceramic	Carp. 20
Zinc Chloride	N	Fluorocarbon Nitrile	Carbon Graphite	Hast. "B"	Ceramic Ceramic Face	Hastelloy
Zinc Cyanide	N	Fluorocarbon Nitrile	Carbon Graphite	316 SS	Stellite Ceramic	SS
Zinc Nitrate	N	Fluorocarbon Nitrile	Carbon Graphite	316 SS	Ceramic	SS
Zinc Oxide	N	Fluorocarbon Nitrile	Carbon Graphite	316 SS	Ceramic	SS
Zinc Phosphate	N	Fluorocarbon Nitrile	Carbon Graphite	316 SS Hastelloy	Ceramic	Hastelloy

TABLE VI

RPM/FPM Conversion Table, Approximate Feet Per Minute

Dia. In.		RPM 500	1000	1500	1750	2000	2500	3000	3600	4000	4500	5000	7000	9000	Circum-ference	Area, Sq. Inch	Dia. In.	
.250	1/4	33	66	99	116	132	165	198	238	264	297	330	462	594	.7854	.0491	.250	1/4
.312	5/16	41	82	123	144	164	205	246	295	328	369	410	574	738	.9817	.0767	.312	5/16
.375	3/8	49	98	147	172	196	245	294	353	392	441	490	686	882	1.1781	.1105	.375	3/8
.438	7/16	58	115	173	201	230	288	345	441	460	518	575	805	1035	1.3745	.1503	.438	7/16
.500	1/2	65	131	196	229	262	327	393	471	524	590	655	916	1178	1.5708	.1964	.500	1/2
.562	9/16	74	147	221	258	294	368	442	530	589	662	736	1030	1325	1.7672	.2485	.562	9/16
.625	5/8	82	164	246	287	328	410	491	590	655	737	819	1147	1474	1.9635	.3068	.625	5/8
.688	11/16	90	180	270	316	361	451	541	649	721	811	902	1262	1623	2.1598	.3712	.688	11/16
.750	3/4	98	196	294	344	392	490	588	707	784	882	980	1374	1766	2.3562	.4418	.750	3/4
.812	13/16	106	213	319	372	425	532	638	766	851	957	1064	1489	1914	2.5525	.5185	.812	13/16
.875	7/8	115	229	344	401	459	573	688	825	917	1032	1147	1605	2064	2.7489	.6013	.875	7/8
.938	15/16	123	246	369	430	492	615	737	885	983	1106	1229	1721	2212	2.9452	.6903	.938	15/16
1.000	1	131	262	392	458	524	655	785	942	1047	1178	1309	1832	2356	3.1416	.7854	1.000	1
1.062	1 1/16	139	278	417	487	556	696	835	1002	1113	1252	1391	1947	2504	3.3379	.8866	1.062	1 1/16
1.125	1 1/8	147	295	442	516	590	737	884	1061	1179	1327	1474	2064	2653	3.5343	.9940	1.125	1 1/8
1.188	1 3/16	156	311	467	545	623	778	934	1121	1245	1401	1557	2179	2802	3.7306	1.1075	1.188	1 3/16
1.250	1 1/4	163	327	490	573	654	817	981	1178	1309	1472	1636	2290	2945	3.9270	1.2272	1.250	1 1/4
1.312	1 5/16	172	344	516	601	687	859	1031	1237	1375	1547	1719	2406	3093	4.1233	1.3530	1.312	1 5/16
1.375	1 3/8	180	360	540	631	721	901	1081	1297	1441	1621	1802	2522	3243	4.3197	1.4849	1.375	1 3/8
1.438	1 7/16	188	377	565	659	754	942	1130	1356	1507	1696	1884	2638	3391	4.5160	1.6230	1.438	1 7/16
1.500	1 1/2	195	393	589	687	785	976	1178	1414	1570	1717	1953	2749	3533	4.7124	1.7671	1.500	1 1/2
1.562	1 9/16	205	409	614	716	818	1023	1228	1473	1637	1841	2046	2864	3683	4.9087	1.9175	1.562	1 9/16
1.625	1 5/8	213	426	639	745	852	1065	1277	1533	1703	1916	2129	2981	3832	5.1051	2.0739	1.625	1 5/8
1.688	1 11/16	221	442	663	774	885	1106	1327	1592	1769	1990	2216	3096	3981	5.3014	2.2365	1.688	1 11/16

1.750	1 3/4	229	458	687	821	916	1145	1374	1649	1832	2061	2290	3207	4117	5.4978	2.4053	1.750	1 3/4
1.812	1 13/16	237	475	712	831	949	1187	1424	1709	1899	2136	2374	3323	4272	5.6941	2.5802	1.812	1 13/16
1.875	1 7/8	246	491	737	860	983	1228	1474	1769	1965	2211	2457	3439	4422	5.8905	2.7612	1.875	1 7/8
1.938	1 15/16	254	508	762	889	1016	1270	1523	1828	2031	2285	2539	3555	4570	6.0868	2.9483	1.938	1 15/16
2.000	2	261	524	785	916	1057	1309	1571	1885	2094	2356	2618	3663	4610	6.2832	3.1416	2.000	2
2.125	2 1/8	278	557	835	974	1114	1392	1670	2004	2227	2506	2784	3898	5011	6.6759	3.5466	2.125	2 1/8
2.250	2 1/4	295	590	884	1032	1179	1474	1769	2122	2358	2653	2948	4127	5306	7.0686	3.9761	2.250	2 1/4
2.375	2 3/8	311	622	933	1089	1245	1556	1867	2240	2489	2800	3112	4356	5601	7.4613	4.4301	2.375	2 3/8
2.500	2 1/2	327	655	982	1145	1309	1636	1963	2356	2618	2945	3271	4579	5890	7.8540	4.9087	2.500	2 1/2
2.625	2 5/8	344	688	1032	1204	1376	1720	2063	2476	2751	3095	3439	4815	6190	8.2467	5.4119	2.625	2 5/8
2.750	2 3/4	360	721	1081	1261	1441	1801	2162	2594	2882	3242	3603	5044	6485	8.6394	5.9396	2.750	2 3/4
2.875	2 7/8	377	753	1130	1318	1507	1883	2260	2712	3013	3390	3767	5273	6780	9.0321	6.4918	2.875	2 7/8
3.000	3	392	785	1178	1374	1571	1962	2355	2827	3141	3533	3925	5495	7060	9.4248	7.0686	3.000	3
3.125	3 1/8	409	819	1228	1433	1638	2047	2456	2948	3275	3685	4094	5732	7369	9.4175	7.6699	3.125	3 1/8
3.250	3 1/4	426	852	1277	1490	1703	2129	2555	3065	3406	3832	4258	5961	7664	10.2102	8.2958	3.250	3 1/4
3.375	3 3/8	442	884	1326	1548	1769	2211	2653	3183	3537	3979	4422	6190	7959	10.6029	8.9462	3.375	3 3/8
3.500	3 1/2	458	916	1374	1604	1833	2290	2749	3299	3663	4121	4579	6420	8240	10.9956	9.6211	3.500	3 1/2
3.625	3 5/8	475	950	1425	1662	1900	2375	2849	3419	3799	4274	4797	6649	8548	11.3883	10.321	3.625	3 5/8
3.750	3 3/4	502	983	1474	1719	1965	2456	2948	3537	3930	4421	4913	6878	8843	11.7810	11.045	3.750	3 3/4
3.875	3 7/8	508	1015	1523	1777	2031	2538	3046	3655	4061	4569	5077	7107	9138	12.1737	11.793	3.875	3 7/8
4.000	4	523	1047	1570	1833	2094	2618	3141	3770	4186	4710	5233	7336	9432	12.5664	12.566	4.000	4

TABLE VII

Degrees Centigrade/Degrees Fahrenheit

Deg C	Deg F	Deg C	Deg F	Deg C	Deg F	Deg C	Deg F	Deg C	Deg F	Deg C	Deg F
−40	−40.0	8	46.4	56	132.8	104	219.2	152	305.6	200	392.0
−39	−38.2	9	48.2	57	134.6	105	221.0	153	307.4	201	393.8
−38	−36.4	10	50.0	58	136.4	106	222.8	154	309.2	202	395.6
−37	−34.6	11	51.8	59	138.2	107	224.6	155	311.0	203	397.4
−36	−32.8	12	53.6	60	140.0	108	226.4	156	312.8	204	399.2
−35	−31.0	13	55.4	61	141.8	109	228.2	157	314.6	205	401.0
−34	−29.2	14	57.2	62	143.6	110	230.0	158	316.4	206	402.8
−33	−27.4	15	59.0	63	145.4	111	231.8	159	318.2	207	404.6
−32	−25.6	16	60.8	64	147.2	112	233.6	160	320.0	208	406.4
−31	−23.8	17	62.6	65	149.0	113	235.4	161	321.8	209	408.2
−30	−22.0	18	64.4	66	150.8	114	237.2	162	323.6	210	410.0
−29	−20.2	19	66.2	67	152.6	115	239.0	163	325.4	211	411.8
−28	−18.4	20	68.0	68	154.4	116	240.8	164	327.2	212	413.6
−27	−16.6	21	69.8	69	156.2	117	242.6	165	329.0	213	415.4
−26	−14.8	22	71.6	70	158.0	118	244.4	166	330.8	214	417.2
−25	−13.0	23	73.4	71	159.8	119	246.2	167	332.6	215	419.0
−24	−11.2	24	75.2	72	161.6	120	248.0	168	334.4	216	420.8
−23	−9.4	25	77.0	73	163.4	121	249.8	169	336.2	217	422.6
−22	−7.6	26	78.8	74	165.2	122	251.6	170	338.0	218	424.4
−21	−5.8	27	80.6	75	167.0	123	253.4	171	339.8	219	426.2
−20	−4.0	28	82.4	76	168.8	124	255.2	172	341.6	220	428.0
−19	−2.2	29	84.2	77	170.6	125	257.0	173	343.4	221	429.8
−18	−0.4	30	86.0	78	172.4	126	258.8	174	345.2	222	431.6
−17	+1.4	31	87.8	79	174.2	127	260.6	175	347.0	223	433.4
−16	3.2	32	89.6	80	176.0	128	262.4	176	348.8	224	435.2
−15	5.0	33	91.4	81	177.8	129	264.2	177	350.6	225	437.0
−14	6.8	34	93.2	82	179.6	130	266.0	178	352.4	226	438.8
−13	8.6	35	95.0	83	181.4	131	267.8	179	354.2	227	440.6
−12	10.4	36	98.8	84	183.2	132	269.6	180	356.0	228	442.4
−11	12.2	37	98.6	85	185.0	133	271.4	181	357.8	229	444.2
−10	14.0	38	100.4	86	186.8	134	273.2	182	359.6	230	446.0
−9	15.8	39	102.2	87	188.6	135	275.0	183	361.4	231	447.8
−8	17.6	40	104.0	88	190.4	136	276.8	184	363.2	232	449.6
−7	19.4	41	105.8	89	192.2	137	278.6	185	365.0	233	451.4
−6	21.2	42	107.6	90	194.0	138	280.4	186	366.8	234	453.2
−5	23.0	43	109.4	91	195.8	139	282.2	187	368.6	235	455.0
−4	24.8	44	111.2	92	197.6	140	284.0	188	370.4	236	456.8
−3	26.6	45	113.0	93	199.4	141	285.8	189	372.2	237	458.6
−2	28.4	46	114.8	94	201.2	142	287.6	190	374.0	238	460.4
−1	30.2	47	116.6	95	203.0	143	289.4	191	375.8	239	462.2
0	32.0	48	118.4	96	204.8	144	291.2	192	377.6	240	464.0
+1	33.8	49	120.2	97	206.6	145	293.0	193	379.4	241	465.8
2	35.6	50	122.0	98	208.4	146	294.8	194	381.2	242	467.6
3	37.4	51	123.8	99	210.2	147	296.6	195	383.0	243	469.4
4	39.2	52	125.6	100	212.0	148	298.4	196	384.8	244	471.2
5	41.0	53	127.4	101	213.8	149	300.2	197	386.6	246	474.8
6	42.8	54	129.2	102	215.6	150	302.0	198	388.4	248	478.4
7	44.6	55	131.0	103	217.4	151	303.8	199	390.2	250	482.0

TABLE VIII

Decimal Equivalents

Fraction	Equivalent	Fraction	Equivalent
1/64	0.015625	33/64	0.515625
1/32	0.03125	17/32	0.53125
3/64	0.046875	35/64	0.546875
1/16	0.0625	9/16	0.5625
5/64	0.078125	37/64	0.578125
3/32	0.09375	19/32	0.59375
7/64	0.109375	39/64	0.609375
1/8	0.1250	5/8	0.6250
9/64	0.14065	41/64	0.640625
5/32	0.15625	21/32	0.65625
11/64	0.171875	43/64	0.671875
3/16	0.1875	11/16	0.6875
13/64	0.203125	45/64	0.703125
7/32	0.21875	23/32	0.71875
15/64	0.234375	47/64	0.734375
1/4	0.2500	3/4	0.7500
17/64	0.265625	49/64	0.765625
9/32	0.28125	25/32	0.78125
19/64	0.296875	51/64	0.796875
5/16	0.3125	13/16	0.8125
21/64	0.328125	53/64	0.828125
11/32	0.34375	27/32	0.84375
23/64	0.359375	55/64	0.859375
3/8	0.3750	7/8	0.8750
25/64	0.390625	57/64	0.890625
13/32	0.40625	29/32	0.90625
27/64	0.421875	59/64	0.921875
7/16	0.4375	15/16	0.9375
29/64	0.453125	61/64	0.953125
15/32	0.46875	31/32	0.96875
31/64	0.484375	63/64	0.984375
1/2	0.500	1	1.0000

GLOSSARY

AXIAL SEAL. See Mechanical Face Seal.

BALANCE, HYDRAULIC OR PNEUMATIC. The mathematical ratio of two areas: usually the area of the sealing face which is bounded by the balance diameter and the inside diameter of the seal face, and the area which is bounded by the outer and inner diameters of the sealing face.

BALANCE, PRESSURE. Syn: hydraulic or pneumatic balance.

CARBONIZATION. A reduction of hydrocarbons, resulting in the formation of carbonaceous residue.

CAVITATION. Formation of gas or vapor bubbles within a liquid stream which occurs where the pressure is reduced to the vapor pressure. When these bubbles collapse suddenly (as this pressure is increased), high local impact pressures are produced, which can contribute to seal damage.

COEFFICIENT OF ELASTICITY. An alternate term for modulus of elasticity.

COLD FLOW. Continued deformation with time of a plastic material subjected to continuous load. A similar phenomenon occurs with metals at elevated temperatures; with many plastics, however, deformation can be significant even at room temperatures or below; thus the name "cold flow."

COMPRESSION SET. The extent to which rubber is permanently deformed by a prolonged compressive load.

CRYOGENIC TEMPERATURES. Those temperatures in the range of −162.6 to −452°F.

DIMENSIONAL STABILITY. The ability of a material to retain precisely the shape in which it was molded, fabricated, or cast.

DRY GAS. Gas with a dew point of −50°F or below, or gas with less than 20 ppm moisture.

END FACE SEAL. See Mechanical Face Seal.

FACE SEAL. See Mechanical Face Seal.

FLASHING. A rapid change in fluid state from liquid to gaseous. This can occur in the dynamic sealing interface of a mechanical face seal due to frictional heat being added to the fluid film as the latter is sheared between the opposing faces, or when fluid pressure is reduced below the fluid's vapor pressure because of a pressure drop across the sealing faces.

FLATNESS. Condition of surface which does not deviate from a plane.

FLATNESS TOLERANCE. Total deviation permitted from a plane. It consists of the distance between two parallel planes within the limits of which the entire surface so toleranced must lie.

FRICTION, DRY. The resistance to motion which exists when a dry, solid object is moved tangentially with respect to the surface of another which it touches or when an attempt is made to produce such motion.

GRAVITY, ABSOLUTE SPECIFIC. The ratio of weight in a vacuum of a given volume of material to weight in a vacuum of an equal volume of gas-free distilled water. Sometimes called simply "specific gravity," but not the same as specific gravity which is based on weight measurements in air. For practical purposes, specific gravity is more commonly used.

GRAVITY, SPECIFIC. The ratio of the mass of a unit volume of a material at a stated temperature to the mass of same volume of gas-free distilled water at a stated temperature.

HARD FACE. A seal face of high hardness, applied to a softer material, by flame-spraying, plasma-spraying, or electroplating; can also be achieved by nitriding, carburizing, or welding.

HARDNESS. The measure of a material's resistance to localized plastic deformation. Most hardness tests involve indentation, but hardness may be reported as resistance to scratching (file test) or the bounce of a falling object from the material (Scleroscope hardness). Some common measures of indentation hardness are Brinell hardness number, Rockwell hardness number, ASTM hardness number, diamond pyramid hardness number, and Durometer hardness. Hardness often is a good indication of tensile and wear properties of a material.

HARDNESS, BRINELL NUMBER (BHN). A measure of the indentation hardness of metals, calculated from the diameter of the permanent impression made by a ball indentor of a specified size pressed into the material by the specified force. BHN increases with increasing indentation hardness.

HARDNESS, DIAMOND PYRAMID NUMBER (DPHN). The measure of the indentation hardness of a material. It is the amount of plastic deformation caused by a 136 deg pyramidal diamond indentor under a specified load. Also known as Vickers hardness.

HARDNESS, DUROMETER. The measure of the indentation hardness of plastics and rubber. It is the extent to which a spring loaded steel indentor protrudes beyond a pressure foot into the material.

HARDNESS, MICROHARDNESS. The hardness of microscopic areas. Microhardness values differentiate hardness of constituents in a material.

HARDNESS, ROCKWELL NUMBER (RHN). The index of indentation hardness measured by a steel ball or diamond cone indentor. RHN is given in various scales (B, C, R, etc.) depending on indentor and scales used.

HARDNESS, ROCKWELL SUPERFICIAL. The measure of surface hardness of thin strip of finished parts on which large test marks cannot be tolerated or shapes that would collapse under normal Rockwell hardness test loads.

HARDNESS, SCLEROSCOPE. The measure of hardness or impact resilience of metals. A diamond-tipped hammer falls freely against specimen from a fixed height and rebound height is measured. Scleroscope hardness is read on an arbitrary scale where 100 represents average rebound from a quenched high carbon steel specimen.

HARDNESS, VICKERS. The alternative term for diamond pyramid hardness.

HEIGHT, FREE. The axial length of a mechanical face seal in its non-compressed state.

HEIGHT, OPERATING. The nominal axial length at which a mechanical face seal will be required to operate.

HEIGHT, SOLID. The axial length of a mechanical face seal when the assembly has been compressed to a relatively solid state as indicated by a sharp increase in the load/deflection rate of the assembly.

HYSTERESIS, MECHANICAL FACE SEAL. The difference in load for the same deflection as read from two load-deflection curves, one obtained when compressing, and the other when relaxing, a mechanical face seal.

INTERFACE. The region between the static and dynamic sealing surfaces in which there is contact, or which experiences the closest approach. It is the surface contact which effects the primary seal.

LOAD-DEFLECTION DIAGRAM. A plot of load vs. corresponding deflection.

LOAD, FACE. The axial load computed as the sum of the hydraulic or pneumatic load and the mechanical load.

LOAD, HYDRAULIC OR PNEUMATIC. In a mechanical face seal, the axial load resulting from fluid-pressure forces only.

LOAD, MECHANICAL. In a mechanical face seal, the axial load ensuring contact between the mating faces, normally supplied in part, or in total, by a mechanical spring device, without regard to fluid pressure.

LOAD, OPERATING. The amount of mechanical force resulting from deflecting a mechanical face seal to its nominal operating height.

LOAD, UNIT FACE. See Pressure, Face.

LUBRICANT. A substance whose function it is to control friction and wear; it may also be useful to control temperature and corrosion, to transmit power (hydraulic), to damp shock (viz., dash pots, gears), to remove contaminants (flushing action), and to form a seal (grease).

LUBRICATION, BOUNDARY. A condition of lubrication in which the friction between two surfaces in relative motion is determined by the properties of the surfaces and by the properties of the lubricant other than viscosity.

LUBRICATION, HYDRODYNAMIC. A system of lubrication in which the shape and relative motion of the sliding surfaces cause the formation of a fluid film having sufficient pressure to separate the surfaces.

MATING RING. A flat smooth ring against which the seal nose runs to effect a seal. Often called seal seat.

MECHANICAL FACE SEAL. A device which seals by virtue of axial contact pressure between two relatively flat surfaces in a plane at right angles to the axis of the shaft.

OPERATING RANGE (AXIAL). The minimum and maximum axial lengths at which a mechanical face seal must operate.

POROSITY. The relative extent of volume of open pores in a material. Ratio of pore volume to overall volume of material in %.

PRESSURE. Force per unit area, usually expressed in pounds per square inch (psi).

PRESSURE, ABSOLUTE. (psia) The sum of atmospheric and gage pressure.

PRESSURE, ATMOSPHERIC. Pressure exerted by the atmosphere at any specific location. (Sea level pressure is approximately 14.7 psia.)

PRESSURE, DIFFERENTIAL. The difference in pressure between any two points of a system or component.

PRESSURE DROP. See Pressure, Differential.

PRESSURE, FACE. The face load divided by the area of the seal face, normally expressed in pounds per square inch (psi).

PRESSURE, GAGE. (psig) Pressure differential above or below atmospheric pressure.

PRESSURE HEAD. The pressure due to the height of a column of body of fluid.

PRESSURE, PROOF. The nondestructive test pressure in excess of the maximum rated operating pressure.

PRESSURE, STATIC. The pressure of a fluid at rest.

PRESSURE, SUCTION. The absolute pressure of the fluid at the inlet of a pump.

PRESSURE, VAPOR. The pressure, at a given fluid temperature, in which the liquid and gaseous phases are in equilibrium.

PV FACTOR. The product of face pressure and relative sliding velocity, usually stated in pounds per square inch and feet per minute. The product is a measure of service severity to be considered in choosing materials of construction.

RATE, LEAKAGE. The quantity of fluid passing through a seal in a given length of time. For gases it is normally expressed in standard cubic feet per minute (SCFM) and for liquids in terms of cubic centimeters (cc) per unit time.

RATE, SEAL. The force required to compress a seal assembly a unit distance, normally expressed in pounds per inch.

RATE, SPRING. The force required to compress a spring a unit distance, normally expressed in pounds per inch.

ROTOR. See Mating Ring.

ROUGHNESS, SURFACE. An arithmetical expression of deviation of the measured profile from a nominal profile. It does not include random defects in the surface being measured.

SEAL, BALANCED. A seal constructed so that under a given set of operating conditions the net hydraulic force acting to open or close the seal is approximately zero. In these designs, the only significant face load acting to close the seal is mechanical, usually applied by a spring device.

SEAL, BELLOWS. A type of mechanical seal which utilizes a bellows for providing secondary sealing.

SEAL FACE. The relatively flat facing of a seal nose.

SEAL HEAD. In a mechanical face seal, the assembly consisting of housing, seal nose, secondary seal and spring device.

SEAL NOSE. The part of the seal head which forms a dynamic seal when run against a mating ring.

SEAL, OVERBALANCED. A seal in which, under a given set of operating conditions, the applied hydraulic pressure, or sealed system pressure, results in a net hydraulic force which tends to close the seal.

SEAL SEAT. See Mating Ring.

SEAL, UNDERBALANCED. A seal in which, under a given set of operating conditions, the applied hydraulic pressure, or sealed system operating pressure, results in a net hydraulic force which tends to open the seal.

SEAL WASHER. See Seal Nose.

SHAFT ECCENTRICITY. The radial distance which the geometric center of a shaft is displaced from the axis of rotation. Equals one-half the shaft runout.

SHAFT RUNOUT. Twice the distance which the center of a shaft is displaced from the axis of rotation; that is twice the eccentricity.

SQUEEZE FILM. The fluid layer formed when a film of viscous fluid is forced out from between a pair of approaching surfaces. Pressures are developed in this process which resist the tendency of the surfaces to come together.

SUBLIMATION. The direct conversion of a substance from solid state to vapor state without passing through a transitory liquid state. The vapor, upon recondensing, reforms into the solid state with no intervening liquid phase.

TEST, ACCELERATED LIFE. Any set of test conditions designed to reproduce in a short time the deteriorating effect obtained under normal service conditions.

TEST, ACCELERATED SERVICE. A service or bench test in which some service condition, such as speed, or temperature, or continuity of operation, is exaggerated in order to obtain a result in shorter time.

TEST, BENCH. A modified service test in which the service conditions are approximated, but the equipment is conventional laboratory equipment and not necessarily identical with that in which the product will be employed.

TEST, LIFE. A laboratory procedure used to determine the amount and duration of resistance of an article to a specific set of destructive forces or conditions.

THERMAL ABSORPTIVITY. The fraction of heat impinging on a body which is absorbed.

THERMAL CONDUCTIVITY. The rate of heat flow in a homogeneous material, under steady conditions, through unit area, per unit temperature gradient in direction perpendicular to area. Thermal conductivity is usually expressed in English units as Btu per square foot per hour per degree Fahrenheit, for a thickness of 1 inch.

THERMAL DIFFUSIVITY. The rate at which temperature diffuses through material. Ratio of thermal conductivity to product of density and specific heat, commonly expressed in sq ft per hr.

THERMAL EXPANSION. The increase in dimensions of a solid, resulting from an increase in temperature. When heated or cooled, materials undergo a reversible change in dimensions which depends on the original size of the body and the temperature range studied. In addition to the change in dimensions, a change in shape may occur; that is, the expansion or contraction may be anisotropic.

THERMAL EXPANSION, CUBICAL COEFFICIENT OF. For any particular material, the ratio of the change in volume to the original volume, per degree temperature change. Unit increase in volume of material per unit rise in temperature over specified temperature. Mean coefficient is mean slope of this curve over specified range of temperature.

THERMAL EXPANSION, LINEAR COEFFICIENT OF. For any particular material, the ratio of the change in length to the original length, per degree temperature change. Unit increase in length per unit rise in temperature over specified temperature-range. Slope of the temperature–dilation curve at a specified temperature. Mean coefficient is mean slope between two specified temperatures.

THROTTLE BUSHING. A close fitting stationary bushing through which the shaft passes and rotates freely. Leakage is controlled by the clearance between the shaft and the bushing. Often made of bronze or filled fluorocarbon resin to minimize friction in cases where the bushing may rub against the shaft because of shaft whip or deflection.

TORR. The unit of pressure used in vacuum measurement. It is equal to 1/760 of a standard atmosphere, and for all practical purposes is equivalent to one millimeter of mercury (mm Hg). Example:

$$25 \text{ mm Hg} = 25 \text{ torr}$$
$$1 \times 10^{-3} \text{ mm Hg} = 10^{-3} \text{ torr } (= 1 \text{ millitorr})$$
$$1 \times 10^{-6} \text{ mm Hg} = 10^{-6} \text{ torr } (= 1 \text{ microtorr})$$

UNITIZED MECHANICAL FACE SEAL. An assembly in which all components necessary for accomplishing the complete sealing function are comprised in a single package.

VACUUM. The term denoting a given space that is occupied by a gas at less than atmospheric pressure. For degrees of vacuum see *Vacuum Level.*

VACUUM LEVEL. The term used to denote the degree of vacuum. (a) Rough vacuum – 760 torr to 1 torr. (b) Medium vacuum – 1 torr to 10^{-3} torr. (c) High vacuum – 10^{-3} torr to 10^{-6} torr. (d) Very high (hard) vacuum – 10^{-6} torr to 10^{-9} torr. (e) Ultra high (ultra hard) vacuum – below 10^{-9} torr.

VISCOSITY. Resistance of a fluid to flow; the evidence of cohesion between the particles of a fluid. It is that physical property of a fluid which causes it to offer a resistance, analogous to friction, to the relative sliding motion of two adjacent particles. This property, chiefly noticeable when the fluid is in motion, is the cause of all so-called fluid friction. Viscosity is ascertained by an instrument termed a viscosimeter (sometimes spelled *viscometer*), of which there are several makes, viz., Saybolt Universal; Tagliabue; Engler (used chiefly in continental countries); Redwood (used in British Isles and Colonies). In the United States the Saybolt and Tagliabue instruments are in general use. With few exceptions, viscosity is expressed as the number of seconds required for a definite volume of fluid under a specified head to flow through a standard aperture at constant temperature.

VISCOSITY, ABSOLUTE. In a fluid, the tangential force on unit area of either of two parallel planes at unit distance apart when space between planes is filled with fluid in question and one of planes moves with unit differential velocity in its own plane.

VISCOSITY COEFFICIENT. The shearing stress necessary to induce a unit velocity flow gradient in a material. The viscosity coefficient of a material is obtained from ratio of shearing stress to shearing rate.

VOLATILIZATION. The transition of either a liquid or a solid directly into the vapor state. In the case of a liquid, this transition is called evaporation whereas in the case of a solid, it is termed sublimation.

WEAR. The undesired removal of material due to mechanical action.

WEAR FACTOR, K. A proportionality factor relating the wear life of a given specimen of a material operating against a specific mating surface at combinations of pressure and velocity below the material's PV limit, normally without any extrinsic lubrication.

ABBREVIATIONS AND PREFIXES

ANSI........American National Standard Institute
ARP..........Aeronautical Recommended Practice
ASA..........American Standards Assn.
ASLE........American Society of Lubrication Engineers
ASME.......American Society of Mechanical Engineers
ASTM.......American Society for Testing Materials
AND.........Air Force-Navy Aeronautical Design Standards
AN ....Army-Navy Aeronautical Standards (Air Force-Navy Aeronautical Standards)
AISI ....American Iron and Steel Institute
API...........American Petroleum Institute
atm...........atmosphere
ASHRAE...American Society of Heating, Refrigerating and Air Conditioning Engineers
C or cent...centigrade
cc............cubic centimeter
cm...........centimeter
cfm..........cubic feet per minute
deg..........degrees
D or dia.....diameter
F or fahr ...Fahrenheit
fpm..........feet per minute
fps...........feet per second
FPS..........Fluid Power Society
ft.............foot
"Hg..........inches mercury
hr............hour
Hz............Hertz (cycles per second)
ICC..........Interstate Commerce Commission
in ....inch
ID...........inside diameter
IPS..........iron pipe size
ISO ....International Organization for Standardization

JIC..........Joint Industry Conference
kg............kilogram
kHz..........thousand Hertz (cycles per second)
lb.............pound
LOX.........liquid oxygen
MS...........military standard
MIL..........military
mm...........millimeter
max..........maximum
min...........minimum
MSS.........Manufacturers Standardization Society of the Valve and Fittings Industry
NFPA.......National Fluid Power Assn.
NMTBA....National Machine Tool Builders Assn.
NPT.........National Std. Pipe-Tapered
NPTF.......National Std. Pipe, Tapered-Fuel (dryseal)
NPSM.......National Std. Pipe, Straight-Mechanical
NPSC.......National Std. Pipe, Straight-Couplings
NPSI........National Std. Pipe, Straight-Internal (dryseal)
NPSL........National Std. Pipe, Straight-Lock Nuts
NPSH.......National Std. Pipe, Straight-Hose Couplings and Nipples
NAS..........National Aerospace Standard
NASA.......National Aeronautics and Space Administration
OD...........outside diameter
psi............pounds per square inch
psia..........pounds per square inch, absolute
psig..........pounds per square inch, gauge
PV............(see Glossary)
QPL.........Military Qualified Products List
r or rad......radius
rpm..........revolutions per minute
RMS.........root mean square
SAE..........Society of Automotive Engineers
SSU..........Saybolt Universal seconds (or SUS)
SSF..........Saybolt Furol seconds (or SFS)
sp gr.........specific gravity
sq ft..........square foot
std............standard
temp.........temperature
vel............velocity
VI............viscosity index

MISCELLANEOUS

Area of circle = diameter squared × .7854

$$\text{Centigrade} = \frac{5 \times (\text{degrees Fahr.} - 32)}{9}$$

Centimeters = inches × 2.54

Circumference = diameter of circle × 3.1416

Cubic centimeters = fluid ounces × 28.4
= cubic inches × 16.3871

Cubic inches = cubic centimeters × .0610237

Diameter of circle = circumference × .3183

$$\text{Fahrenheit} = \frac{9 \times \text{degrees Cent.}}{5} + 32$$

Feet per minute = diameter (inches) × rpm × .2618

Inches = centimeters × .3937
= millimeters × .03937
= mils × .001

Inches of Hg = psi × 2.04

Kilograms = pounds × .453592

Kilograms per sq mm = pounds per square inch × .000703

Millimeters = inches × 25.4
= mils × .0254

Millimeters of Hg = pounds per sq in × 51.7

Mils = inches × 1,000

Pounds = C wt × 112
= Kilograms × 2.2046

Pounds per sq in = Kilograms per sq mm × 1422.3
= inches of Hg × .49
= millimeters of Hg × 0.0193
= feet of water × .43

Square centimeters = square inches × 6.4516

Square inches = square centimeters × 0.155

INDEX

ABOUT THE AUTHOR

A holder of numerous mechanical face seal design patents, John C. Dahlheimer's previously published work on the subject appeared in *Machine Design* magazine. He has participated in corporate technical exchange programs with European manufacturers of such seals, and frequently holds mechanical face seal seminars for users throughout this country. He attended the University of Illinois and Northwestern University prior to holding product engineering positions with two diverse mechanical face seal companies in the Chicago area.

Dahlheimer joined International Packings Corporation, Bristol, New Hampshire, in 1965. After establishing capabilities relative to these types of seals, he now administers their Mechanical Face Seal Department. Although primarily responsible for design suitability and product integrity, as Product Manager he remains deeply involved in research and development, methods of production and quality assurance, marketing, field technical assistance and other activities related to these sealing devices.

Dahlheimer owns a home in the beautiful lakes region area of Laconia, New Hampshire, where he, his wife and three sons enjoy their favorite recreational activity of boating on the family's cruiser.